MW00510288

Memories & Recipes

by

Janet Stuart Smith

Southern/French Recipes

Published by Janet Stuart Smith
P.O. Box 11114
Memphis, TN 38111
Fax: 901-458-2306
www.lunaweb.com/janet
jsmith@lunaweb.net

LIBRARY OF CONGRESS CATALOGING-IN-PUBLICATION DATA

Justine's: Memories and Recipes

Summary
ISBN # 0-9667324-0-5

Book Design and Typography: Gloria C. White & Associates

Paintings by Janet Smith

Acknowledgements

Thank you to all my friends for their encouragement,
motivation, proofreading and organizing

...and especially to Wilma Madison for writing the recipes from her heart.

Contents

Dayton and Justine Smith

I would like to dedicate this book
to my father, my ghost writer,
who would have loved
to see his name all over the place.
And to my mother, who worked from
morning until midnight for 48 years.
They were a partying twosome who
definitely had the lust for life.
They never slowed down.

"Said by many prominent Memphians, the biggest attraction downtown was Justine's, Memphis' most venerable restaurant.

Sir Rudolf Bing described Justine's as an "oasis" on the Metropolitan Opera's annual tour.

It is the restaurant to which society has been going for years, and its lure predated the revival of downtown."

– *Town and Country*
November, 1982

Preface

A neighbor of mine died recently who was well endeared for his chili that he would share with the neighborhood friends. It was an all-day-or-more cooking affair, where everything was put together just right. I asked his son, by chance, did they ever write the recipe down. "No," he replied, it was gone forever.

I remember my father at home would also make an "all day cooking affair" with all the fresh vegetables grown out of our garden, with who-knows-what thrown in. The soup was duly named "Fire-Hole" soup and my friends and I would enjoy lunch together, always looking forward to the next batch. It was something, well, just taken for granted, I guess, and now, since my father has died, I wish, among other things, that I had written his recipe down. After thinking all of this through, I realized that what I had really taken for granted, all of my life, were the recipes I had surrounding me from Justine's, my parents' restaurant. So, here are the recipes and my memories...the memories of growing up in a very successful restaurant, and an era and a restaurant that, I believe, will be hard to replace.

Janet Smith

Introduction

Thanks to my mother's innate culinary expertise, my father's management and intense devotion to authentic restoration along with the same excellent staff for four decades, Memphians, their guests, and welcome travelers dined in elegance befitting the term "Southern Hospitality."

No one person, besides the original founder Justine herself, can take all of the credit for the success story of Justine's; it was, most assuredly, a group effort to which I hope to make tribute in this book.

Wilma Madison, one of the many great chefs at our restaurant, cooked the dishes for years and recently put the recipes down for me to write this book and, of course, for Justine herself to proofread.

Wilma Madison, 1959, her first day

My mother had never written any recipes down, always tasting and measuring in the palm of her hand, changing until it was just right; so, needless to say, it was very entertaining for her to see them in print.

No one could ever have had a better upbringing than I had, growing up at Justine's, with all the wonderful "family" of employees. Every night seemed like opening night, a new adventure, a new start. Life was never routine. Something new and unexpected was always happening throughout the 48 years.

My parents each had other families before they married each other and had me. Dad called me his "do-it-yourself" grandchild. It was fun having nieces and nephews my age or older.

My parents accomplished so much in their forties. Now that I have reached that magical age, I remind myself that success can be reached at any point in our lives.

My family

Thanks

Wilma Madison and J. D. Hill.

Most employees last at a restaurant for only a year or two while the average length of employment for forty-six employees at Justine's was at least thirty years. We all shared such a feeling of friendship and camaraderie at work. We were not the typical "restaurant types." Since their hours were not as long as the kitchen staff's, many of the waiters were moonlighters with other prominent jobs. Some even held master's degrees. I believe we had a very classy group of employees.

Many of the waiters were trained by Ervin McDonald, the head waiter, who worked there for thirty-four years until his legs finally gave out. Ervin was the most famous and irreplaceable stately southern maitre d'. He greeted the customers by name in his black tuxedo and bow tie as they entered the building.

Ernest Gillespie *Ervin McDonald*

Lorita Ward

Jan Nolen, Mike Cannon, Maudie Walker, Elizabeth Smith, Robert Webber, Wilma Madison, Beverly Campbell

My parents were fortunate to have the same group of devoted, hard-working, "six-nights-a-week" employees for so many years. These people gave up time with their own families because there were very few days off. With the exception of Christmas and Thanksgiving, and a two-week vacation at the end of every August, most of their lives were spent at the restaurant.

The employees got it done and they got it done right every night. Justine's was famous for its consistency. Always the best, the freshest, nothing else....

Julia Taylor

The restaurant was such a part of our lives that I knew if the time ever came, I would miss it and everyone I worked with there. It was such a big wonderful "family" of talented, hard-working employees alongside the nicest group of people you would ever want to meet. I couldn't leave them and we all couldn't leave each other. It was definitely a group effort.

From top: Robert Miller, Lamar Walker, L.T. Walls, Julius Winselle, Melvin Mackey, Boston Brown, Charles Conley, Henry Crossley, Sr., Anderson Bridges, Jan Nolen, Dayton Smith, Ervin McDonald, James Stanback, Maurice Taylor, Jr.

Life consisted of long hours, long weeks and challenges every evening to do the best we could. Our group effort was definitely a labor of love. With my mother's skill of choosing the freshest ingredients, the cooks' magic touch, and the organizational expertise of my Dad, everything ran to near perfection. Everyone was always working at a heavy pace and never had any time to just hang around. The evenings went by so quickly that at the end of every night, we felt as if we had accomplished "Mission Impossible."

Maudie Walker, Melvin Mackey, Ethel Anderson, Ervin McDonald

List of Employees

Jimmy Adams
Joe Allen
Ethel Anderson
Ridley Anderson
Rene Ball
Neddie Banks
Jake Barber
Earl Bogan
Anderson Bridges
Boston Brown
Willie Mae Brown
Richard "Rip" Burns
Beverly Campbell
Willie Campbell
Mike Cannon
Charles Conley
LaShuan Cotton
Rodney Cotton
Henry Crossley, Sr.
Henry Crossley, Jr.
Floyd Cunningham
Robert Cunningham
James Dailey
John Davis
Kramer Davis
Alan Dehamandi
Frances Finnie
Jim Getten
Ernest "Bunky" Gillespie
Willie Gray
Virginia Greer
Charles Gregory, Sr.
Charles Gregory, Jr.
Lula Mae Gunn
H.B. Hall
Paul Harding
James "Cass" Harris
Joe Harris
Joe Henderson
James D. "JD" Hill
Sylvester Hill
Norma Holcombe
Mary Nell Horner
Robert Houston
Willie B. Hughes
Charles Jones
Clemmie Jones
Ronnie Jones
Ajean "Boogie" Karnes
Andy Khalihan
Alice Kimball
George and Lillian "Tiny" Kolb
James Knight
Alice Faye Lee
Pinky Loveless
James Lundy
Pamela Madison
Sharon Madison
Wilma Madison
Mary Malone
Rose Malone
Melvin Mackey
Dennis Maki
Bryan Martin
Ervin McDonald
Charles "Jackie" Means
Willie Merit
Howard Myers
Andre Miller
Reginald Miller
Robert Miller
Robert Montgomery
Alice Morck
Leo Neal
Darrell Nolan
Jan Nolen
Ken Nolen
Pearl Odom
R.G. Ogden
Jack Perkins
Lajoy Phillips
Genie Pique
Herbert Powers
Ernest Purnell
Bill Rainey
George Redmond
Elizabeth Rightor
Frank Robinson
Howard Robinson
Martin Robinson
Amanda Roland
Patricia Sadler
Carl Sanders
Robert Scales
Serdina
Raymond Shaw
Billy Sights
Louise Skiles
Georgia Smelcer
Carl Smith
Elizabeth Smith
James Stanback
Wilton Steinberg
Charles Sueing
James Arthur Tatum
Julia Taylor
Margaret Taylor
Maurice Taylor, Sr.
Maurice Taylor, Jr.
Helen Thompson
Frances Walker
Lamar Walker
Maudie Walker
Melvin Wallace
L.T. Walls
Donny Ward
Lorita Ward
Orlander Warmsley
Quan Watson
Robert Webber
Jack Westbrook
Joe Westbrook
Richard Williams
Sylvester Williams
Webster Williamson
Julius Winselle
Sam Wolfe

Beginning

I was going to name my book "Justine's – The Agony and the Ecstasy" or, for a better description, "Bistro 911." All I know is, in 1948 my mother Justine started it all in an old warehouse in Memphis, Tennessee. The location was near what was called the city car barn and the famous Beale Street. The first bottle club in Memphis, Th' Sharecropper, was right next door where people would drink while waiting for their table. The address was 242 Walnut, which never was and never will be a nice part of town. All that remains of the address now, sadly, is a chain-link fenced-in parking lot.

With no practical experience in business at all, she quietly opened. She claims she was broke, playing poker, mah-jong and bridge, and needed some way to make money.

She swore her family to secrecy and did not tell her friends, just in case it was a failure, wanting to wait until it was up and running. She brought all of her nice linens and Baccarat crystal from home and with great luck and hard work, Justine's quickly became the "in" place for entertaining and dining. Customers drinking and having fun ended up throwing her crystal against the walls. She would smile, secretly aware of who would have the last laugh. She knew it would all work out in the end because she was going to serve the finest food in this part of the country, no matter what the cost.

Within a month of the opening, *The Washington Post* carried a favorable review. The writer's cab driver let him off at the Walnut Street address warning him that it was not a safe neighborhood. He went in anyway and wrote a beautiful article. Within two years the restaurant was receiving recognition in the national media and later, I am very proud to say, became one of the leading restaurants in the United States, lasting 48 years.

In 1955, my mom married my father, Dayton. Nine months to the day later, they had me, their love/perfect child. Daddy (a Yankee, thus traitor, my mother would say, "because he liked to keep a clean house") always said he could divorce my beautiful mom at any time because she married him under false pretenses. She claimed that she was in her 30's when they married, when, in fact, she delivered me when she was 42.

My Parents

My father was the son of W. D. and Bessie Smith of Dundee, Illinois, one of four brothers. He had two children by a previous marriage named Dayton and Patricia. My mother was the daughter of George and Bessie Holdway of Newport, Tennessee, one of eight children. She also had two children by a previous marriage named Justine and Dan.

Dad was a Chicago businessman, entrepreneur and newcomer to Memphis when he first met, and soon married, my mother, in 1955. He was President of The Worth Products Co., whose address was 884 Linden, located around the corner from the original Justine's.

He said he really intended to take her to Chicago, but he liked the roots she had in Memphis, and decided to get into the restaurant business. They were married in New Orleans.

“Dayton had thousands of friends in Memphis and around the country,” said restaurant owner John Grisanti. “I think he was just a little more than a restaurant owner. He was good for our community.”

Dad was an engineer, researcher and restorer.

“A bon vivant with a good business head, Dayton Smith has been the most important person in Justine’s personal and business life.”

–Alice Fulbright
The Commercial Appeal

It was fun growing up around my parents’ endless activities: working in the vegetable garden, riding my horse “Red,” taking care of our home and all of Justine’s.

Essential Ingredients

What you need in the kitchen:

Two Hands
The Freshest Ingredients
Our Staff

And, believe me, you don't need any fancy equipment to prepare any of these dishes. I do not remember anything in that kitchen that the cooks could use to make life any easier for them in preparing these dishes every night, for all of those years.

Everything was prepared from scratch every day; the sauces, the dressings, all of the ingredients. And, surprisingly, there were never any leftovers.

Fresh was the rule.

Just attention to detail from people who really had a remarkable touch and took pride in what they were doing.

In the daytime, it was just a small depressing no-nothing kitchen, but at night, it was magically transformed into New Orleans or Paris, where miracles were performed and masterpieces were created and where you wondered, how did they do it?

Here are the complete, never-written-down-until-now recipes. This is what they made every day in proportions suitable for the home.

It was fun dressing up and joining the event, which was the business every night. The food, the atmosphere and the production were certainly one-of-a-kind where many generations enjoyed dining and celebrating special events.

Fortunately for me, since my parents had me in their forties, I believe I was given a lot more freedom than most children early on. Since they had "been there and done that" so to speak, they turned me loose in the restaurant, where I could always find something interesting to watch or get my hands into.

The Menu

Preston Battistella of Battistella's Seafood in New Orleans, was the long-time supplier of Justine's and a large part of our success. Mr. Battistella would send us the freshest seafood available each day before selling to his Mardi Gras customers. I remember Mother, after spending long nights at the restaurant, would get up at six o'clock every morning to call in the order. My parents were the first to use the airlines to fly the seafood in every day which was delivered by a fast cab to the restaurant.

Of course, you could adjust these recipes to your caloric intake if it would make you feel any better; it wouldn't be any fun, but you get my drift.

I gained weight after Justine's closed.

NOUS RECOMMANDONS

Huitres en coquille a la Rockefeller 1.50
Cocktail aux Crevettes 1.00
Crevettes remoulade 1.00
Roties de Champignons Sous Cloche 1.50
Harengs en Daube ..... .85
Anchois sur canape ..... .85
Crabes sur canape Justines 1.75
Cocktail aux Crabes 1.25
Escargots a la Bourguignonne 1.50
Shrimp de Jonghe 1.50

POTAGES

Soupe a l' oignon gratinee ..... .50
Vichyssoise ..... .50
Bisque de Homand (commander d' avance) 2.50
Bisque de Pomparts Normande (commander d' avance) 2.50
Gombo Creole ..... .60

Menu from the 1960's

The front door at Justine's

The Evening

For an evening at Justine's, you would get all dressed up, "coat and tie for the gentleman," would be greeted at your car by Jackie and at the Williamsburg Blue front door by Ervin McDonald who, most always, knew your name and where you wanted to be seated.

Hopefully and shortly thereafter, your waiter would appear, Mackey, Anderson, Henry, Conley...to take your cocktail order. You then would sit back and soak up the ambience. The antiques, the tall ceilings, the paintings....

Your cocktail would arrive with a basket full of hot buttered French bread toast that you would find hard not to eat too much of before dinner. You would watch the customers come in the front door through the heavy red velvet curtains and be escorted to their tables. Looking around you would recognize other tables of guests and the party would begin.

"Got to go visit so and so, here comes whatchamacallit...excuse me, I need to go upstairs to the powder room"...up the grand staircase...The waiter would return, seeing if you needed another cocktail and would proceed to hand you the tall, dark green, slick, vinyl-covered menu full of drawings by Billy Price Hosmer Carroll, who used to take me along with her, when I was a kid, while teaching at the art academy.

You would look through the four pages of menu items trying to organize in your mind what to order. "Should I start with my favorite appetizer or entree or.... I can't eat all of that..but, I want all of that...they are my favorites...they are all my favorites...." There were over 100 selections.

There was not a bad selection on that menu, it was perfect. The only bad thing was deciding what to eat that night, not regretting ordering something else, and, most of all, walking out the door, unaided and comfortably, when you left.

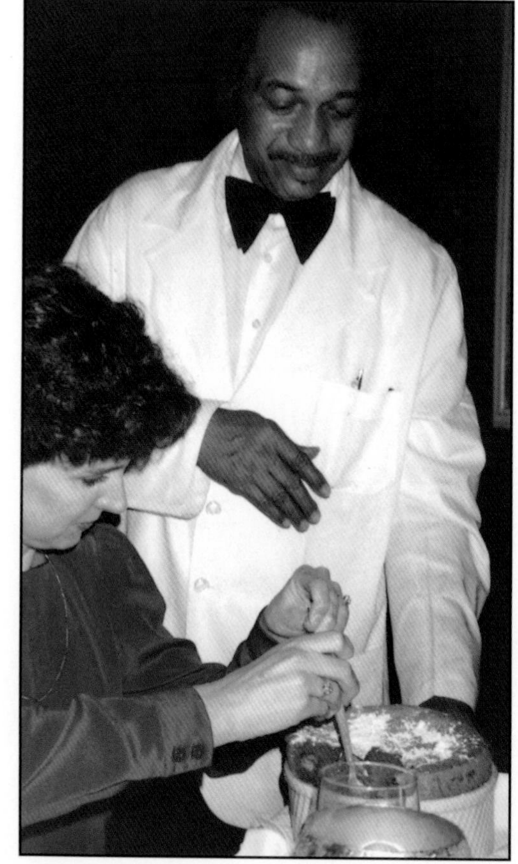

Our wine list was the best. What wine to order with what dish or what dish to order to go with that special bottle of wine...there were many choices. When you finally decided and closed your menu, running over the plan in your mind, thinking of any last minute changes, hoping the waiter would walk more slowly to your table, you sat up in your chair, grabbed your water or cocktail and grinned from ear to ear feeling a sense of accomplishment.

Memphis restaurants refining their wine service

In 1970, when liquor by the drink was legalized in Memphis, the first Memphis restaurant to put in an extensive wine list was Justine's.

One of the wines the owner-hostess of the restaurant bought in the early years, after wine became legal for restaurants to serve, was the now fabled Beaulieu Vineyards Georges de Latour Private Reserve Cabernet Sauvignon 1970 Vintage, which she bought in half bottles. In this vintage, it is one of the world's great red wines, and Justine's 150-case pine-binned wine cellar still contains about three cases.

Other prize bottles listed include a Charles la Franc Maison Rouge 1957, a Chambertin Clos de Beze 1961, a Chateau Lafite Rothschild 1964, a Georges de Latour Private Reserve 1958, and a Chateau Mission Haut Brion 1953. Browsing through the bins, I also discovered a 1954 Lafite-Rothschild, a 1959 Chassagne Montrachet, and a 1921 Grand Vin de Leoville.

Justine's was the first Memphis restaurant to list the "Big Four" Bordeaux – Chateau Lafite Rothschild, Chateau Margaux, Chateau Latour and Chateau Haut-Brion. A measure of Memphis restaurants' growing sophistication in the service of wines is that, where only about 20 restaurants were serving wine in 1970, around 200 serve it now. Justine's still has one of the finest wine lists among Memphis restaurants, but limits it to 60 selections.

– Edwin Howard
"Life at the Top"
Memphis Business Journal

Surprisingly, the majority of the time, the waiter knew you and knew exactly what you were going to order and you thought "HEY, maybe I'll change next time." And you knew there was going to be a next time! But, you usually saved the next time, as every time, for some special occasion and then you would just make coming out special, deciding you were special and you did not need a reason. The place just made you feel good. Being there made you feel special, put you on show, made you feel glamourous, sophisticated.....

Walnut Address

I don't believe many people from my generation know there was a Justine's on Walnut. I wish I could remember the old place, not that I wish I was born 10 years earlier, I just wish someone had taken some photographs. My parents never filled me in on the first years. There was always too much going on so, unfortunately, I never asked.

I do know that in the 40's, during the war, there was a shortage of projectile rings (whatever those were). They were made in a warehouse on Walnut owned by Ira Allstadt. Ira, who would later marry one of Mom's sisters, needed a cafeteria to feed his employees, so Mom, ready to earn some extra cash, made sandwiches for him.

When the war ended and the cafeteria was no longer needed, Ira told Mom she could use his equipment and set up a restaurant for herself in the corner. During those years, the best places for entertaining and dining in Memphis were in one's own home or private club. The leading restaurant in the city featured T-bone steaks and French-fried onion rings served with something less than loving care. She saw a situation not becoming to a city steeped in the traditions of Southern hospitality and elegance. So, determined to learn as she went along, Mom thought she would give it a try.

where gourmets gather

JUSTINE'S

unexcelled French and Southern cuisine served in an atmosphere of leisure and hospitality

272 Walnut

For reservations phone 37-5744

First ad – 1948

But, first of all, she wondered what she should call the place. She worried about using her own name in case it was a failure and her friends would know, but not coming up with anything better, in November of 1948, at 35 years old, on her first marriage with two kids, Mom quietly opened "Justine's."

She worked from eight in the morning until midnight (as until the end) and flew to New Orleans on her only off day to observe operations at Brennan's and Antoine's, where she discovered the coffee she would use. Operating on a shoestring budget and using the second-hand kitchen equipment, she made her own draperies and employed two cooks and three waiters. The renowned Melvin Mackey was the first to walk in the door, even before she had opened, and stayed until the end, 48 years later.

On opening night she invited ten friends to come by for dinner. The ten friends came, and so did twenty others who had heard of her venture and within a matter of weeks she was having capacity crowds every night in the eighty-six seat restaurant she had named Justine's. There was little advertising except word-of-mouth and not even a sign out front.

Nous Recommandons

Huitres en coquille à' la Rockefeller	1.00	Anchois sur canapé	.85
Cocktail aûx crevettes	.75	Canape Lorenzo	1.00
Crevettes rémoulade	.75	Crabes sur canapé Justines	1.00
Champignons frâis sur toste	1.00	Cocktail aux crabes	.75

Potages

Soupe à l'oignon gratinée	.40	Gombo Créole	.50
Vichyssoise	.40	Potage tortue au sherry	.50

Entrees

Tournedos Beauharnais	$ 4.00
Tournedos à la Chartres	3.85
Filet de boeuf béarnaise	4.00
Filet de boeuf aux champignons frais	3.75
Châteaubriand	6.50
Faison en cocotte (commander d'avance)	4.75
Filet de boeuf Robspierre en Casserole	4.50
Pigeonneaux sauce paradis (commander d'avance)	3.85
Poulet sauce Rochambeau	2.50
Poulet en cocotte	2.50
Poulet sauce Florentine	2.25
Poulet aux champignons frais	2.25
Poulet à la crapaudine	2.00
Langouste a la Gresham	2.50
Poulet a la creole	2.00
Oeuf Sardou	1.50
Brochettes de Foies de Volailles	2.50
Coquilles Saint Jacques au Gratin	1.85
Coquilles Saint Jacques au Cari	2.00
Coquilles Saint Jacques à la Marinière	1.85
Coquilles Saint Jacques en Brochette	1.85
Cassola de Peix	2.50
Cuisses de Grenouilles à l'Osborn	2.25
Crevettes à la Créole	2.00
Crevettes à la Clemenceau	1.90
Pompano à la Claudet	2.50
Pompano amandine	2.75
Pompano grillé	2.50
Pompano en papillote	2.50
Flounder Louisiane	2.50
Flounder à la Mariniere	1.90
Filet de truite à la Marguery	2.25
Filet de truite amandine	2.25
Filet de truite florentine	2.25

Excerpt from first menu – 1948: Minimum per customer $3.00

After two months she was able to pay off her loans and, exhausted, closed the doors and went away to recuperate.

Excerpt from an old Commercial Appeal society column:

Sunday night, thanks to the hospitality of an open-handed epicurean, we joined a group dining at Justine's. Glancing around at her fellow diners, Penelope espied Pat and Miller Jameson, Pat in tasteful navy, supping with the Henry Jacksons. Richard Maury and fiancee, Frances McCain, were enjoying a "tete-a-tete" as they awaited their orders. Frances' silk print was most becoming. At another table were Jean and Bill Wills, Jean suited in oh-so-vogueish gray, Sam and Ann Portlock, her navy frock accented by a spanking white choir-girl collar, and Shep and Doug Kenworthy. The Portlocks, sad to relate, are soon moving to Georgia; in fact, Sam has already opened his Chevrolet Agency in Douglasville and will shortly be joined by Ann and the children, so the gathering Sunday night, which had earlier included Jeanne and Jimmy Owen and Jack and Imogene Erb, was by way of a temporary farewell. Penelope says temporary for she hopes the Portlocks will be buzzing back and forth to Memphis as often as possible.

Bobo Rockefeller didn't come here.

Report that Bobo Rockefeller, wealthy estranged wife of Winthrop Rockefeller of Little Rock, was in Memphis dining with a bachelor, proved false yesterday.

The bachelor, identified by diners at a swank restaurant on Walnut, Saturday night, said the woman confused with Mrs. Rockefeller was a secretary of Mr. Rockefeller. He presumed diners overheard the name "Rockefeller" and drew the wrong conclusions. The secretaries, two of them, dined with the bachelor, came to see *South Pacific.*

A New York spokesman for Mrs. Rockefeller, said flatly, she was not in Memphis.

– *Memphis Press Scimitar*
No date

I later learned the "bachelor" was my father.

My parents flew down to New Orleans one night and were married. Mom met the Brennan sisters through their brother Owen, with whom she used to have cocktails after closing hours in their restaurant, when she was "learning the ropes." Adelaide Brennan became one of my godmothers. Restaurant owners became "family."

Seating on Walnut was in one row along the left side. The window seat was for the V.I.P.s.

That was the beginning, and for years later, Justine's was recognized as one of the leading restaurants in the United States.

Back in those days, everyone in Memphis was selling coffee for five cents a cup. Mom charged ten.

My parents loved to dance. In the early days, after work each night, they would go to the Skyway Room at the Peabody Hotel. Come to think of it, even in the later years, I think they must have lived to dance after closing the restaurant each evening. They began with ballroom dancing and went to disco. Where that energy came from, I don't know.

Some very tasty prices on the 1948 Justine's Menu.

Let's just figure the bill for my favorite five-course dinner at Justine's: Oysters Rockefeller, $1; watercress and potato soup, 50¢; watercress and bacon salad $1; Pompano a la Claudet, $1.90; Justine's Special Dessert 75¢ – oh yes, and coffee, 10¢.

That comes to $5.25, plus tax and tip.

Those were 1948 prices, the year the restaurant opened on its original site at 272 Walnut Street.

– Edwin Howard
"Life at the Top"
Memphis Business Journal
October 5-9, 1987

Gay Nineties Apartment

My parents lived miles out in East Memphis when White Station was a gravel road. They developed an apartment, as a convenience to their business, over Th' Sharecropper next door. It was a good place to drop in and rest, entertain friends or sleep overnight if necessary. Here is where Mom first saw my father's restoration capabilities.

Dad used the "Gay Nineties" theme to decorate the apartment. The point was, nothing should be new. It ended up being a treasure house of late 19th Century Americana. The quarters featured every type of memento such as mahogany louvered shutters, art glass, gas jet chandeliers, candlelight, cut glass, noisy floral wallpaper, heavy fringed velvet drapery and overstuffed furniture. It had the nostalgic flavor of the tight-laced corset complete with a perambulator, a bowl shade lamp and a beaded portiere.

When my father couldn't find what he wanted, he made it. A hanging beaded screen separating the hall from the bar was the "Smith" version of the old-time glass portiere which shimmered between living and dining rooms in many an old home.

Dad manufactured his own portiere from colored wooden beads and commercial tubing. He hung it from an antique spool design archway, which he painted gold.

Here began the theme of showy grandeur. The room behind the screen was closed in on the street side by heavy burgundy velvet drapery. The opposite wall was a solid mahogany mass of louvered shutters that came out of Memphis' old Gayoso Hotel and which my

father estimated were made in 1850. The only relief was in the brilliant art glass windows lit to show the designs. My father said the room was never meant to be lit. It was actually more of a cocktail lounge, reminiscent of the old hotel bar. The "Smith" bar was a carved walnut fireplace, converted to the purpose. Overhead was another of the softly lit art glass windows, with a hanging lamp of Venetian glass, circa 1757.

There was candlelight from a cast iron Gothic chandelier which held four large candles. Hanging heavily by a crude chain, this fixture imparted a distinctly medieval flavor. In the parlor, where a Gay Nineties living room would have been, there was a brassy, feminine air. Bold rose bouquets on the walls, a gilt mirrored mantel, and an overstuffed wall bench in burgundy velvet under a huge pier mirror marked the room for the era.

Dad's pride and joy was an ornate chandelier which, in another age, burned five gas jets. He wired it for electricity. There was gilt in the desk and in the standing bird cage next to the mantel. Dad said he had to have a bird in a gilded cage.

– Information from an article by Ruth Jacquemine, The Commercial Appeal, June 9, 1957

Coward Place. Opening Night, 1958

Coward Place

Ten years after my mother opened her restaurant on Walnut Street, my father decided it was time to move into surroundings more befitting her classic cuisine. In 1956, he fell in love with a stately old plantation house, the Coward Place, located just around the corner from Walnut Street. It was one of the earliest examples of Italianate architecture in the Mid-South.

The property was first deeded, in 1818, to Andrew Jackson, John Overton and James Winchester. Around 1834, Nathaniel Anderson bought the land and, with slave labor, built his home. The herringbone walks in front of the building were made from clay dug from the carving out of the basement. The same clay was used in the bricks that made up the 18-inch thick walls of the building.

Anderson sold the building for $3,000 to cotton planter H.W. Grosvenor, who enlarged the house. Grosvenor wanted to emulate the French colonial architecture he had seen in a similar house during a steamboat trip to New Orleans.

In 1866, the house was sold to the Cowards whose offspring, Memphis' Russell and Austin families, turned over a wealth of mementos and trivia to my family during the restoration.

"The Coward Place" stood at the end of East Street before that street was built. Lamar Avenue, then a plank road, ran past the north boundary of the property. Mr. Coward's obituary, in the 1800's, referred to the location as "near the southeastern outskirts of the city," where it took hours, by carriage, to get to Front Street.

During the Civil War period, the Coward Place had been the scene for debut parties for the society belles. Bullets blew through the windows during the war and musket balls were left embedded in the 18-inch thick walls. There was a beautiful winding staircase, ornate frescoed moldings, French mirrors and massive furniture.

When my father bought the building, it had been turned into apartments and the front was a battleship gray. Aided by Maude Cawthon, as technical consultant, he searched out and even dug out the past. He studied the building from every angle, happily pointing out the one remaining original window which he matched with all the others. He ripped away other additions, made by the previous owners, particularly at the rear. Dad said the house had been remodeled at least six times. He said he just peeled it back like an onion to its original state.

Dad spent fourteen months restoring the building back to its original 1834 stature, just as it was at the height of the Victorian Era. In the beginning, people thought he was crazy. After he stripped off seven coats of paint and got down to the original pink stonecote, bystanders called it "Dayton's pink elephant."

A large Venetian porch at the entrance was removed when a foundation study revealed the size of the original entrance. Maude and my father deduced that the porch was added during a later period and was not part of the original building. Dad found small pegged holes in the front of the structure which indicated that a small porch with an ironwork grill around the top was there first. The original front marble steps, which led up to the small entrance porch of old black and white marble, were found buried in the back yard.

Other wrought iron was brought over from the old Hewett place located on Vance Avenue.

The warm pink exterior, found under the coats of paint, was the original stonecote, made with baked crushed limestone, red sand and brick dust. Underneath the stonecote, they found the countless handmade bricks made long

Photo by L. Uoykcuf

The Garden

ago. The kiln was discovered on the grounds. A dark green growth of thick English ivy would cover the front of the building years later.

The old fashioned garden was where some of the finest old trees in Memphis were teemed with boxwoods, magnolias and scores of other plants. My father loved to point out an ancient pine which still leans markedly to the east as having been a meeting place for the Chickasaw Indians. The garden patio was formed of large stones which came from Jefferson Davis' home on Court.

One part of the courtyard was reached by an immense wrought-iron gate that had been in Memphis since 1881, with indications of a much older Deep South history. The gate swung between huge brick piers.

A driveway of hand made brick was laid in the popular herringbone pattern and ran down the side of the building through carriage gates. One could imagine a surrey rolling toward the old carriage house at the rear.

The carriage house was restored but later destroyed by lightning. It had been complete with a tin roof, pecky cypress board and batten walls with antique hardware. Later, the boards would fence in part of the back yard. I learned, early on, the meaning of "recycling." A hand-hewn cypress gate was nearby, not far from the summer outdoor kitchen which held a huge brick oven. During cocktail parties, Willie, J.D. or Boogie could not seem to crack the oysters open fast enough for the customers, who would suck them down by the handfuls around the outdoor pit.

Outdoors, each evening, shafts of light illuminated the original cornices of the structure.

The garden walls, gates and grounds shaped up nicely during the restoration, but the interior work was slower. It took more time to gather the furnishings to make the place a showcase once more.

The spacious 14-foot-tall rooms had once hosted many memorable social events. The west dining room had a beautiful elaborate fresco around the ceiling.

After shopping trips to New Orleans for antiques, French clocks, iron work and statuary, the new location became a museum full of romance and charm. The entrance hall became the foyer of Justine's with magnificent brass candelabra and with walls covered with white paper in which a gold chandelier design was repeated. Two magnificent original Venetian Blackamoor statues guarded the black and white foyer. Each of the antique chandeliers in the dining rooms had an individual historic background. The east room chandelier once graced the home of Napolean's general, Maréchal Murat.

My father used the balustrade from the renowned Gayoso Hotel in Memphis to make the banister of the entrance hall's colonial stairway. Italian marble lavatories were recycled from the hotel as was a cashier cage which was made into a garden gate. All of this and much more were complemented by treasures of antique Georgian silver and Baccarat crystal.

Being the genius that he was, Dad arranged the kitchen at "zero level" where the doors would open and the air would rush in, leaving no food odors in the house.

Sentiment played a part in the remodeling. One dining room was an exact replica of a room that was in Justine's on Walnut. Draperies there were the deep red, recycled velvet and French murals were on the walls. Candlelight flooded the room very softly.

My father always said that the forerunner of this room, which we called "The Justine's Room," was where Memphis couples first courted and later returned, in its new location, to celebrate wedding anniversaries. The room was always there for these customers to remember and reminisce.

Dad took great pride in doing something that would be representative of early Memphis. He was proud of what the restaurant had done for the neighborhood formerly known as Suzette Bottoms.

An ancient octagonal gazebo in the garden would, years later, inspire the pavilions in the rear of the garden, with their eight-sided theme. An octagonal rose garden was also built which boasted 47 varieties of beautiful roses such as Sonya, First Prize, Swathmore and Peace.

Beyond the rose garden was a glassed-in piano bar with the same function as Th' Sharecropper, that had been next door to the first restaurant. Guests could have a drink while awaiting their tables. The garden recaptured the romantic era of the 19th century and offered a beautiful setting to complement the cuisine that had brought Justine's its renown. It was the perfect architectural and botanical setting for the lovely old mansion.

The Gazebo in the 1960's

Two years before my father died, he succeeded in having Justine's included on the National Register of Historic Places. It was mainly to ensure against future condemnation, preserving the building as a Southern haven. Dad always took care of business. He never left any stone unturned.

As many as four generations, including some who actually lived in or were entertained at the antebellum home, could be found dining with us on any evening.

The history of the Coward Place and times were so interesting. Without my father, it may have never been preserved. I hope writing this book will be my way of preserving and making tribute to the past, to the accomplishments of my parents and all of the wonderful people that made it happen.

Justine's was long a Memphis landmark.

The Old Coward Place

by Samuel Evan Ragland

1943

In the days of yore in a sylvan wood
A mansion majestic quietly stood;
A place of culture and of classic lore;
A shield of dignity it proudly wore.

Its master noble, a man of learning,
For nature's mysteries ever yearning;
The fields he loved, and the things that grew,
The creatures that walked, and the birds that flew.

The master has gone, and so have the trees,
Save a few whose boughs still stir in the breeze.
The city has moved away from the town
And encompassed the place all round and round.

And the stately mansion looks down a street
Where once loving eyes a forest would meet,
But despite the stress of weather and time
The house is grand with its portal and vine.

And the guest who enter finds warmth and cheer,
Instinctively knows refinement is here.
Worthy the daughter and honor'd the sire
Though Coward in name, no craven in ire.

Beyond the house in a sequestered place,
The garden, fringed with ivy lace,
Lies bleak, and drab, and unaware
Of the precious things that sleep in its care.

The flowers are withered, the leaves are sere;
The sky is overcast, the world is drear;
The fountain is dry, and birds have flown,
But they all will return when winter is blown.

Ghost

I had always heard ghost stories about the Coward Place among the cooks and the waiters and always hoped there might be something to it. I had also heard about the waiters' poker games in the wine cellar, nights after another closing, when my parents thought everyone had locked up and gone home. The partakers were really hiding out in the shadows, waiting for my parents to leave so

they could start their game and their night of fun. I heard they played these games many times until one night when the ghost appeared in front of the whole table. That was the last card game.

I wanted to believe there was a ghost but I never had any real reason to until one night when I was the maitre d'. I had to go down into the wine cellar to reset the air conditioner. It was dark and quiet, and as I crept down to open an old wooden door, where the reset button was located, I felt someone behind me. Looking on the side of the door, I saw a tall dark shadow. I thought it was one of the employees trying to scare me. When it didn't go away and no one answered me, I turned around to to see a figure. She was in a hazy gray gown, recessing back towards the restrooms, half there and half not, and her feet were not touching the ground. I did not tell anyone for a long time, wanting to revel in the experience and not to be told I was crazy.

On another night, years later, I went upstairs to check on a party and found one of the customers sitting out in the hall looking dazed. I thought "Oh, no, she's sick or something..." and asked if she needed anything. She looked right up at me and quickly said, "Is there a ghost story to this house?" I said "yes" and started to go into my spiel about "Aunt Mary, who stays in the wine cellar, looking for her child, who died in childbirth, during the Civil War...." She interrupted saying, "No, she's upstairs, in this hall, something is going back and forth up here and over to that room!" pointing to the ladies room. "I can really feel it, she's active, upset...." I believed and reassured her, knowing that my encounter was even more credible. I recently learned the upstairs ladies room was a nursery in the Civil War days.

Bathrooms

Even the bathrooms were famous. On June 5, 1994, Lisa Jennings wrote in *The Commercial Appeal*:

"In the ladies' restroom upstairs, an elegant window seat reflected in a large mirror. The room used to be a nursery and where many a bride got dressed. It was a spot for a moment's peace and privacy, a place to rest [and for me to hide from the downstairs hall full of customers]. Sunsets streamed through curtained bay windows onto a black and white tiled floor and polished marble sinks.

Delicate boudoir-style chairs sat welcoming preeners to a mirror that covered half the wall. In a small room off to one side, two toilets sat behind slatted doors. In the evenings, there were fresh roses in the restrooms. An old photo on the wall showed the garden's gazebo years ago, looking the trend at the time.

The ladies' room upstairs had an air of dignified elegance, though the original floor tiles were cracked and chipped.

There was a "unisex" restroom off the restaurant's foyer. Here, a massive carved marble mirror framed the small room. The marble sink came from the old Gayoso Hotel. On the walls hung articles on the restaurant from *The Atlanta Journal Constitution*, *The Memphis Business Journal* and *The New York Times* in which Craig Claiborne called Justine's 'conceivably, the best restaurant in the South.' Over the toilet, sadly, hung an etched portrait [that I did] of the late Joe Henderson, a member of the waiter staff for many years."

Joe was notorious for driving up to work and forgetting to put his car in park.

Awards

Justine's won its share of awards. We were often listed in business surveys as one of the top 100 restaurants in the country and appeared on "best" lists many times over.

Within ten years of its original opening, Justine's began appearing regularly on the prestigious *Holiday* magazine list of distinguished restaurants.

For 22 consecutive years, Justine's received the celebrated Holiday Award as one of the nation's top dining establishments. The magazine recognized the finest restaurants in North America, eschewing the ordinary and unimaginative. We outlasted the Holiday Awards.

"The awards carried their sincere endorsement of those dedicated establishments that provided dining experiences beyond the commonplace. Their choice of restaurants, often landmarks in their communities, were dependably excellent and unequivocally honest in serving an enthusiastic patronage. The awards were based upon criteria concerned with the selection, preparation, and service of foods, the atmosphere, and the attentive, sensitive sincerity of performance. Value became more and more important when it came to spending dollars of one's discretionary income."

– *Holiday Magazine*

Parties

"Fun and frivolity of Memphis 'happenings,' combined with worthy philanthropic causes, were given or hosted at Justine's, as when the pavilions formally opened in 1978. The occasion, that evening, was a black tie dinner-dance for the Kidney Foundation. Everybody was excited about the opening of the outdoor pavilions. Dayton had said for years that he wanted to build the gazebos, reminiscent of the Victorian era." [Dad fell off the roof earlier in the day and was unable to attend the party.]

– Carol Vaiden
Memphis Press Scimitar
September 18, 1978

The Pavilions

Metropolitan Opera Parties

For 18 years, my parents hosted a glamorous party for the Metropolitan Opera stars and patrons. The traditional late-night dinner followed the Monday night performances in Memphis. Most of the well-known stars have been our guests, from Beverly Sills, and Luciano Pavarotti to Sherrill Milnes. Sir Rudolph Bing, the Metropolitan's manager, announced that the Met would not miss coming to Memphis if for no other reason than to eat at Justine's. He wrote in his book, *5000 Nights at the Opera,* that Justine's was the best restaurant in the United States.

During the parties, you would find Mal and Ernie, violinists from the Lester Lanin Duo from New York, strolling through the crowd, playing popular tunes on their violins. There were magnificent roses.

Mal and Ernie entertain guests.

Guests all had opera stars at their tables. The party was always a complete sellout. My parents assumed the full expense of the party where the guest list had to be held at 200

tickets. The proceeds were donated to the Arts Appreciation Foundation, where $100 per person invitations were on a strictly first come, first served basis. The tickets were limited to contributors and their out-of-town guests.

The Met's ballet department was not included in the party at Justine's. My parents wanted them also to receive credit while they were in Memphis, so they would throw a party for them on the last day of their performance. The parties were in the outdoor pavilions where there was plenty of champagne and food and always an orchestra for them to dance to. Those were the best parties. Mike Cannon, our popular piano player, was always there along with the sumptuous food in the silver compotes. We danced and partied until the dancers would leave for their performance at the Auditorium.

Second Night's production, *Madama Butterfly*, had all of the glittering touches of the opening night – beautiful gowns, furs and jewels, and an excitement that comes once a year theatrically, when the Metropolitan Opera plays in town. "Little Janet Smith" was hearing opera for the second straight night. She arrived with her parents and with Mrs. Betty Alexander who accompanied William Pope of Washington, Georgia.

– The Commercial Appeal, *May 8, 1963*

Letter from William R. Kent to my parents

Dear Dayton and Justine,

I am sure a letter from me isn't necessary to tell you what a beautiful evening you gave the Metropolitan Opera company and the Patrons and Patronesses of the Memphis Opera Benefit Association, who sponsored it, as I am sure you have heard from all sides and especially from those for whom it was given.

Our house guest, Gabor Carrelli, couldn't get over it when Rudolf Bing rushed up to me Tuesday night and thanked me for the beautiful party which was far and away the best party on the tour. Bing is noted for disliking social events and apparently always refers to them as something "I must attend and how I dread it." I must say I don't think I have ever seen people enjoy themselves more or eat more.

Everything was perfection, and I only want to say how very, very much we appreciated everything you did to make it such a memorable evening,...including your luck with the weather. Many thanks, and with all best to you both, I am
Sincerely, Bill

– William R. Kent
May 10, 1963

No wonder Memphis received plaudits from the opera greats. Preparations for the opera party given at Justine's required one whole week of preparation, while running the "sold-out" business every night.

Everything was perfection. Even the one crisis that developed. A special order for lobster didn't arrive from the airport till late, but was served. Food served at the party was not only fit for a king, but of course, for the opera stars. The menu that evening: Boeuf Wellington Sauce Perigueux, Blinis au Caviar, Artichokes stuffed with Crabmeat, Sirloin of Beef, Seafood Curry, Lobster Thermaidor, Silver Champagne Bowls filled with Fresh Fruit and numerous other desserts.

– Memphis Press Scimitar, Wednesday, May 8, 1963

Opera finale: Elegant Dinner...for several hundred, the first night at the opera was capped with an invitation to dinner at Justine's. Dinner-dance afterward at Justine's provided a fitting end to a gala evening. The party offered the high-spirited gaiety only associated with a debut. "I'm sorry it's a little cold out here–there's nothing I can do about it," whispered genial host Dayton Smith, as he whirled a dance partner to the tunes of the popular orchestra of Jim Johnson. Dancing [my parents' favorite passion] was outdoors, under a tent, adjacent to the formally landscaped patio and garden area. For those who chose to mingle outside, there was much to keep them warm. Hot Oysters Rockefeller were being dipped off the grill as soon as they were done, while platters of fresh oysters were constantly replenished nearby. Under the tent were other hors d'oeuvres, hot seafood casserole and pâté.

Justino Diaz, the Met singer who played the role of Colline, was one of the first party guests to arrive. He wore a casual, beige leisure suit and sunglasses and appeared to be just a bit taller than he had seemed onstage.

Vincente Sardinero, *Boheme*'s Marcello, exerted the same bouncy energy, quick smile and seeming friendliness which he had used in the role. At the party, Vincente wore a suit with an open shirt collar, much in contrast to the other tuxedoed male guests, and several necklaces and medallions at his throat. "I'm very glad to be here–we appreciate this party," he said.

Inside, the stars of the show, Pilar Lorengar and Giorgio Casellato-Lamberti were surrounded by Memphis guests.

– *Memphis Press-Scimitar, May 10, 1977*

Once again, Dad's roses, in all shades of red, pink, white and coral, were used in large arrangements in several spots inside the restaurant and also served as centerpieces for each table. There were silver candlesticks and roses galore.

Operagoers knock out a rainy night with gaiety. After the opening night performance of *Thais*, a crowd of 250 adjourned to Justine's where Dayton and Justine hosted a party for patrons and guarantors...

Special guests at the dinner were the stars of the opera. Beverly Sills was the first to arrive, stripped of her on-stage makeup and Egyptian costume. She was dressed in a rust-colored pant suit. Sherrill Milnes arrived a little bit later, outfitted in a tuxedo with a multicolored brocade and velvet vest. Another star, Raymond Gibbs was there and during dinner broke into a charming rendition of a popular song to the accompaniment of Mal and Ernie, who play each year for the party.

Francis Robinson, the Met tour director and consultant, was there spending time renewing acquaintances and making new friends. The executive director Anthony A. Bliss was there along with Charles Rieckers and John Pritchard, the conductor.

– *Memphis Press Scimitar, Tuesday, May 9, 1978*

Symphony Ball

On a Saturday evening in October, the Memphis Symphony Ball was held at the restaurant. The cocktail music that evening was by that popular duo, Mal Burke and Ernie Gallet who always played for the Metropolitan Opera parties we hosted. Michael Carney played out in the pavilions and kept the guests dancing until the wee hours of the morning.

The menu that night was Crabmeat Justine, limestone [fancy name for ice-berg lettuce] and watercress salad, Tournedos Bearnaise, parsley potatoes, broccoli hollandaise and Dessert Justine's, all which were misspelled in the catalog.

Vincent de Frank, the distinguished music director and conductor, attended along with other prominent local names and devoted customers.

Ginger Rogers stopped in around two in the morning. She had been appear-ing in the Cole Porter musical *Anything Goes*, at the Orpheum Theater, and said she was sorry that she missed the music because she would have liked to have danced with and for the guests. She remained at the party until it broke up after 4 a.m.

> Chilly weather, evening dresses, coats over, garden pavilions. Inside, Mal Burke, on guitar, and Ernie Gallet, on accordion, strolled through the dining rooms, stopping at tables to play selections, including "Misty," "Talk of the Town" and "Tomorrow" for the diners. While they were eating, the guests would stop and clap to the beat of the tune.
>
> *–The Commercial Appeal, September, 1978*

Birthday Parties

Mine were always at Justine's.

Justine's had its share of well-known guests....

Photographs by Frank Braden

Arkansas gubernatorial candidate Bill Clinton (circa 1977)

Tastevin Dinners

In 1961, Justine's served as the first host for Memphis' Chevaliers du Tastevin Society, a group of wine connoisseurs.

How did they eat and drink it all??!!

Les Vins et Les Escriteaux

On Boira:
LE CHABLIS, GRAND CRU 1955
VALMUR
LE MOËT ET CHANDON
CHAMPAGNE CUVÉE DOM PÉRIGNON 1952

Réception:
LES BOULETTES DIABLÉES
LES HUÎTRES FUMÉES
LES PETITES CREVETTES AU BEURRE

Première Assiette:
LE CONSOMMÉ DOUBLE

On Boira:
LE BATARD MONTRACHET 1955

Deuxième Assiette:
LES CRÊPES ÉCREVISSE ET HOMARD

On Boira:
LE CHAMBOLLE-MUSIGNY PREMIÈRE CRU 1953 EN MAGNUM
VIGNES DU CHÂTEAU
BERNARD GRIVELET
TASTEVINAGE

Troisième Assiette:
LE CANARD SAUVAGE DE LA RIVIÈRE L'ANGUILLE

Pour se Rafraîcher:
LE TROU DE MILIEU:
CALVADOS ARC DE TRIOMPHE

On Boira:
LE CHAMBERTIN 1949
MAISON J^H FAIVELEY
TASTEVINAGE

Quatrième Assiette:
LE FILET DE BOEUF À LA JUSTINE
LES COURGETTES FARCIES

On Boira:
LE NUITS ST. GEORGE 1955
CLOS DE LA MARÉCHALE
MAISON J^H FAIVELEY
TASTEVINAGE

Issue de Table:
LE PLATEAU DE FROMAGES

On Boira:
TAITTINGER
BLANC DE BLANCS 1953

Boutehors:
LA GLACE AU JULEP DE MENTHE

On Dégustera:
DOM BÉNÉDICTINE
LE CHARTREUSE JAUNE
LA FINE CHAMPAGNE, XO-HENNESSY

LE CAFÉ NOIR TRÈS CHAUD

Chaine des Rotisseurs

We also hosted the first Chaine des Rotisseurs party in Memphis. The Chaine was the food society where Tastevin was for wine. Paul Spitler, a good friend of my parents, was the maestro and the party was held in our wine cellar for years. Mother remembers providing two cases of Chateau d'Yquem 1955 for the dinner.

Confrérie de la Chaîne
des
Rôtisseurs

Chapitre de la Memphis
Diner d' Intronisation

en l'honneur de
M. JEAN VALBY
Paris, France
Grand Chancelier de la
Confrérie de la
CHAÎNE DES RÔTISSEURS

Mardi
Le 20 Fevrier 1962

Atlanta Justine's

OKAY, so after the Walnut Street opening and closing and the move to the Coward Place, after the restoration alongside capacity crowds, feeding and entertaining them, then going dancing after work each evening till the wee hours alongside running and maintaining a ten-acre lot of home and gardens on Rich Road, my parents decided to build in Atlanta. Then, they took up skiing and tennis.

My parents were never content with established success. Five years after the Coward Place restoration and move, my father was restless for more. SO, he attacked another monumental job, another Justine's in a museum setting in Atlanta.

> "It is, of course, unwise to stand in the way of progress as it moves forward and leaves in its wake older and less efficient ways of doing things. Yet, there is a certain beauty about the past, and man is also unwise if he looks only where he is going and forsakes where he has been."
>
> *– From Atlanta Justine's brochure by Paul Hemphill*

Dad had hoped to undertake a restoration similar to the one in Memphis, but he quickly realized that the perfect structure for a museum restaurant – one which was not only representative of Georgia Piedmont architecture but also had the proper location and adaptability for a restaurant – had simply not survived Sherman, and the great Atlanta fire of 1917 and subsequent progress.

He would have to find an attractive piece of land and then transport a house, wherever it might be, to the land. What he finally found, after scouting nearly every known rare "survival" in Georgia, was the Pope Plantation near Washington, Georgia, in historic Wilkes County, 117 miles east of Atlanta.

The plantation had been built in 1797, with a unique "dog trot," or breezeway, bisecting the house, and when my father discovered it at dusk one day, it was virtually obscured by trees and undergrowth. He found cattle using the dog trot for shelter.

Although it had been vacant for fifteen years and was in a state of general disrepair, Dad fell in love with it. "It didn't have the columns of the Greek revival you see so much of," he said. "It was primitive, true Georgia Piedmont architecture."

Pope Plantation

But it was love at first sight, and excavation began immediately. The house would be moved to Atlanta and placed on a lovely hatchet-shaped wooded lot of four acres on Piedmont Road.

With plenty of non-filtered Pall Mall cigarettes that he never inhaled, and Old Fitzgerald, 100 proof, Bottled in Bond, which he did inhale, my father moved the building, piece by piece, to Atlanta, 117 miles away, where it was restored and intentionally hidden from view on a historic tract on Atlanta's north side. The address was 3109 Piedmont Road N.E., located in Buckhead.

The work that followed staggers the imagination. Hundreds of photographs, sketches and measurements had to be made of the old plantation, along with every casement, joint and rafter, before it could be touched. Then, the plantation was dismantled and cataloged, brick by brick and board by board, to be hauled by truck to Atlanta. Five other structures of the same period, from a coach house in Putnam County to a house in Elberton, were located and stripped for supplemental lumber and brick. Some 280,000 brick were required. All of this going on while everyone back in Memphis was working hard keeping the successful, popular, one-of-a-kind, restaurant going.

Dad, Maude Cawthon, Jim Getten

Dad acted as his own contractor, ran his own forge, along with his carpenters, arc welders and cutters, who were all specially trained. Whenever possible, the original materials and hand tools were used in the restoration. There were 180,000 handmade bricks in the complex; one chimney alone used 14,000 bricks. All the floors were the original virgin Georgia pine, hand planed, or hand adzed by eighteenth century craftsmen. The doors in the ladies' room were 160 years old. Only square-cut nails or dowels and hand cut moldings were used. The modern kitchen was added, as an extension of the original building, but it too was completely covered with the same eighteenth century building materials.

The whole complex would take another year to complete and would include, as separate buildings, a cotton gin, well house with the original windlass, spring house, dovecote, privy, and a typical early Georgia forest bridge over a stream running through the property.

The actual restoration took more than three years. It was three years of searching for one more 18th century door, studying Georgia Piedmont architecture, answering the questions of disbelieving passers-by, forging hardware for the massive entry gate with the same facilities used in the late 1700's, and discarding the friendly suggestions of those who knew short-cuts. Wooden pegs and original

lumber were used even in the attic and inner walls, which no one would see. All along, attention had to be paid to the wiring and other special needs of a public restaurant.

When the massive entry gate was finally put into place outside the property, it prompted an Atlanta newspaper to ask under its front page picture of the gate, "When, O When, Will These Great Gates Swing Open?

Those who knew of the renowned Justine's in Memphis were eagerly awaiting the Atlanta counterpart, but as the years passed and nothing happened, rumors grew and the skeptics were taking bets the restaurant would never open. Open it did but, very quietly, with a whisper instead of a shout.

There was no advertisement, publicity or even a name out in front. If you didn't know exactly where you were going, you'd miss it; and yet from the first day, it had been next to impossible to get a reservation. That fact alone made it the talk of Atlanta.

Apart from the food, there was another reason for the Memphis Justine's fame and the Atlanta one would no doubt follow suit. Both restaurants were virtual museums of past eras.

Jean Dubuc, Atlanta chef

Atlanta Justine's

And so, the Atlanta Justine's opened in 1966, half museum and half restaurant, with no promotion except by word of mouth. Nestled in the trees, virtually isolated from the most vibrant city in the South, it stood as a monument of the past, the perfect frame for another fine old painting. Every artifact furnishing the rooms was of the period where you could find the largest collection of George III Sheffield silver in the United States, outside of a legitimate museum.

The funny thing was that we had the same customers in Atlanta that we did in Memphis. People who ate with us were from all over the world and traveled a lot.

The Justine's in Atlanta won almost as many awards as its Memphis predecessor. It was also named a top award winner in *Holiday Magazine.*

From 1966 to 1972, the Atlanta Justine's was in operation. Both of my parents had a tremendous amount of energy, but personally supervising in two cities was exhausting, so they sold the Atlanta restaurant. The routine of meeting each other "coming and going" was too much. (I figured they sold it knowing I would attend college nearby in 1973.)

It would take most of a lifetime to accomplish what my parents did in any given year. They never slowed down until each reached 82.

Justine's in Memphis AND Atlanta each became something of a civic institution.

Information from the Atlanta Justine's brochure
by Mr. Paul Hemphill, 1967.
The Commercial Appeal, November 5, 1962.
Atlanta, "Town Talk," September 1966

Atlanta Justine's Wine Cellar

The Atlanta Justine's, as the Memphis counterpart, specialized in fine beef and fresh Gulf seafood. On the Atlanta menu, one could select from:

Oysters Rockefeller	$2.00
Beluga Caviar	$5.00
Crabmeat Justine's	$2.00
Fresh Maine Lobster	$6.00
Poulet Justine's	$3.75
Crème Spinach	60¢
Chateau Lafite Rothschild	$ 9.00
Chateau Mouton Rothschild	$14.50
Moet et Chandon	$10.00
Dom Perignon	$15.00
Mumms, Extra Dry	$ 9.50

Ending

My father died in 1988 and my mother retired in 1996. A customer wrote: "Justine's provided for Memphians, for so many years, the ultimate in fine dining, wonderful food, and ambiance. It has been a part of people's lives that will never be forgotten."

The closing of Justine's was truly the end of an era. As I said, it was an era and a restaurant that, I believe, will be hard to replace.

Excerpts from Articles

Interesting Lincoln Continental Owners

Clearly, Dayton and Justine Smith are a husband and wife team whose standards in cars (each drives a Continental) are on the same high plane as their standards in business (they own two of the finest restaurants in the South).

Dayton resembles Spencer Tracy and some of his friends think Hollywood lost out because, apparently, a career in acting didn't occur to him. When he married Justine and retired from a career of management engineer, he and Justine became one of the finest restaurant twosomes in the country.

– The Continental Magazine, Fall, 1966

Fortunately, menus and chefs never changed much at Justine's.

Menus and chefs don't change often enough at Justine's, Memphis' pioneer fine-dining restaurant, to require the several reviews a year I may write of other local restaurants. But I do feel a duty to do a quality check of Justine's about once a year. As they say, somebody's got to do it. I eat there much more often than that, of course, for the sheer pleasure of it.

I watched Alexander Gudunov eat three filets, one right after the other, in Justines, several years ago. [At an opera party night, I remember Pavarotti ate two of everything he ordered.] All three pompano dishes were absolutely perfect, the freshest of fish, flown in from the Gulf that day. Our dinner guests were from New Orleans and pronounced the pompano Louisianne, stuffed with crabmeat and mushrooms, far superior to similar dishes in The Crescent City's restaurants.

The filet was fork tender [thanks to JD], perfectly cooked, richly flavorful, with a superb Bearnaise.

The crabmeat Justine and the oysters were all flawless, in the best Justine's traditions. Having these Creole dishes at Justine's, after a recent visit to [Memphis'] Owen Brennan's, which is attempting to offer both Creole and cajun cuisines, offers a study in contrasts that is wholly in Justine's favor. Yet Justine's prices remain among the lowest in town for fine dining.

Next time you order any of the pompano dishes at Justine's, try the Albert Pic Chablis with it. It's a marriage made in heaven.

Justine's is also Memphis' prime party place whether outdoors in the lush gardens, the trio of charming interlocking pavilions and/or the piano bar, or inside on tables handsomely set with family silver. Dixie Carter and Hal Holbrook's party there recently, honoring a nephew's engagement, beautifully utilized both the outdoor and indoor settings, the former blessed by a gentle breeze that wafted the music of a string trio gently on the twilit air.

– Edwin Howard
Memphis Business Journal, 1989

Dayton found a special kind of brick design from William Faulkner's home in Mississippi, that combines the herringbone and basket weave. He copied it for the walk.

Justine and Dayton are a fascinating team and they think big. [BIG! Not just a couple of roses, we ended up with 800 plants at our home and restaurant, counting the garden at the piano bar.] Their formula has made Justine's a unique restaurant. This year *Cosmopolitan* magazine wrote up Justine as one of the country's most successful women in the restaurant business. *Better Homes and Gardens* has featured her, and the restaurant is regularly listed on *Holiday*'s distinguished restaurant list.

Justine is a girl from the mountains of East Tennessee, who created a restaurant in Memphis that is a legend. She has hosted opera stars who come to Memphis each May, famous celebrities, and native Memphians who all agree that her restaurant is in a class of its own. Her food is expensive, distinguished and memorable.

– *Chattanooga News-Free Press*
Wednesday, July 7, 1976 .

For so many years, Justine's was practically the only game in town when it came to fine dining. There were other good restaurants, but none of them had quite the grand presence that Justine's exuded. The sweeping entry foyer, the live piano music, the long stemmed roses, the elegant gardens...all the elements for a special evening were there.

Aside from the sheer pomp and circumstance of the place, Justine's food is the real attraction. The menu is probably the most extensive in town. One could eat there for weeks on end without duplicating a dinner.

–Tom Martin
Memphis Magazine
July, 1983

The elegant old home, around the corner from the Elmwood Cemetery off Crump Boulevard, hosts a black and white tile entryway. White table cloths, dimly lit rooms, the feeling of ghosts of employees and diners past, fill the empty spaces, chatter and clatter of forks against dishes. The rooms seem to swell with the presence of the days-gone-by. Justine is beautiful... femme fatale à la Veronica Lake, time has softened that image into one of great good health and a kind of ageless and robust sweetness. (Ha) she laughs in a slightly fey way. The kind of old age that everyone hopes for: skis, dances, swims, and tennis.

Customers from around the country return time and time again drawn to the culinary and atmospheric seduction.

– Mary Ann Eagle
Memphis Magazine
February, 1994

Cousin Amanda with me and my parents

The Old Elegance Remains Justine's in Memphis

Anyone who passes through the lofty double doors of Justine's restaurant in Memphis is in for a double treat. First, they'll enjoy some of the finest food on the face of the earth. Second, they'll be stepping into a past that exists today mainly in movie sets and historical novels.

The entry foyer from the staircase

Justine's occupies a French colonial mansion that was built in 1834. Eight years ago, when Justine and Dayton were seeking new quarters for their popular restaurant they chose the old mansion. It had had several owners and uncounted remodelings during its 130 years of existence. The Smiths set out to restore it as nearly as they could to its original condition. That they have succeeded remarkably well is a tribute to their taste and unrelenting pursuit of authenticity.

The Smiths combed New Orleans for ornamental iron work, for French clocks and statuary. Magnificent mirrors, paintings and period wallpaper, ornate moldings and fourteen foot ceilings give an atmosphere of elegance. In the East Room guests dine by the light of a chandelier owned by one of Napoleon's generals.

On the grounds, walks of hand-made brick, laid in the old herringbone fashion, meander through the garden areas of boxwood, cherry laurel and magnolias. The massive wrought-iron gate goes back in Memphis history to 1881, or earlier.

Entering Justine's, it is easy to imagine how it must have been in the old days: the glittering, genteel parties, the belles and beaux, the stately music and outside the muffled clopping of horses arriving in the driveway.

Justine's today is one of Memphis' favorite restaurants. There is a $3 per person minimum charge.

– Dodge News Magazine
May, 1965

West Dining Room

Justine's, A Past with a Present

"The reputation they've held for all these years is incredible. If there's anything that's hard to do in the restaurant business, it's to maintain your reputation," says restaurateur John Grisanti.

The restaurant received its most recent honor in January with almost habitual inclusion in a national survey of businessmen as one of the top 100 restaurants in the country. The survey was among subscribers to two New York magazines, *Sales and Marketing Management* and *Restaurant Business.*

A night like this one: Friday, so busy the last few calls for reservations get apologies instead of the pompano or crab. For the rest, candles flicker in a 144-year-old room already moody from a dimmed chandelier and the scent of roses.

Surrounded by dark green wallpaper with gold leaf, guests who need the unchanging 103-item menu strain past the French in large print to microscopic English translations. "Crevettes" for shrimp, "huitres" for oysters and "escargots" for the brave.

There are no microwaves or steam tables in the kitchen – nor deep-fry pots or grease. "You have to push all the time [and have plenty of Hennessey on hand!]."

– *Mid-South Magazine*
The Commercial Appeal
Sunday, July 12, 1987

Justine's endures as a favorite for French cuisine

Among longtime Memphis residents, a handful of restaurants in town will always be enduring favorites. Justine's is one of those places, so revered that the mere mention of the name brings a chorus of swoons from those who have been visiting the grand mansion in gourmet dining for decades.

Inside, Justine's is a dark and romantic hideaway: tables spread with linen, silver and glass, along with small vases of fresh flowers and flickering hurricane lamps.

A tender beef filet was cooked to order – rosy in the center. Confirming a colleague's glowing report based on her dinner there a month earlier, Justine's serves some of the best red meat in the city. This filet was among the choicest cuts we've had anywhere in Memphis.

Of course, we had to have dessert, and the baked Alaska we ordered was the perfect finale to a memorable meal. Premium vanilla ice cream was enveloped beneath a frothy layer of meringue that had then been baked to a toasty brown. Our server deftly spooned rum and brandy over this confection, torched it table side, and stood attentively by as we and diners throughout the entire room watched the blue orange flames lap up the mountainous egg white peaks. After the flames had died, we dove into the gooey, melting mess to find a combination of lush, cool sweetness from the ice cream mixed with the penetrating flavors of the rum and brandy.

"Cass"

Service was pretty much perfect. Our waiter was gracious, personable and put us immediately at ease. Throughout the course of our 90 minute meal, he seemed attentive, ready and waiting to replenish water glasses, refill drinks, discreetly scrape the bread crumbs from our tablecloth, and in general make sure that everything about our meal was just the way we wanted it.

Happily, it was. All in all, our dinner at Justine's was as delectable as it was memorable–which is important, considering we chose the restaurant, as do many other patrons, as the place to celebrate an important birthday.

– Linda Romine
Memphis Business Journal
1995

What becomes this legend most? Rich memories of elegant meals

Forty-eight years is a long time, a nostalgic time. – Justine

Justine's defined fine dining in Memphis. And even when newer, more progressive establishments emerged, loyal customers came to the restaurant for such signature dishes as Crabmeat Justine and Pompano Claudette.

The surrounding neighborhood deteriorated drastically over the years. Typical Memphis urban blight. The restaurant's first charge account was opened by Shelby Foote, who at the time was a budding novelist. William Faulkner always brought his daughter, Jill, to the restaurant when she was traveling to St. Louis to school, and Tennessee Williams brought his grandfather to Justine's for the old man's birthdays. [Mom remembered the last one was his 95th.]

For 18 years, the Smiths hosted dinners for the traveling company of the Metropolitan Opera of New York: in his autobiography, *5,000 Nights at the Opera* (1972), impresario Rudolph Bing called Justine's the best restaurant in the country. Though that estimate might be laid at the feet of enthusiasm, Justine's won its share of awards.

Holiday magazine gave the restaurant top honors for 22 years. Loyal patrons seemed content with Justine's reliance on butter and eggs and the traditional panoply of rich sauces that never disappeared from the restaurant's menu. "The customers have been wonderful, so many of them have been with me since the beginning."

She said her tenure as a restaurateur was finished. She needed to spend time with her family that she hadn't ever really had time to say hello and goodbye to.

– Fredric Koeppel
The Commercial Appeal

Upstairs Dining Room

Other notable firsts, which came to Memphis thanks to Justine's, include the arrival of Muzak, (hers was the first business to pipe it in) and fully trained waiters (most from Alonzo Locke's Waiters School). Some members of her original wait staff were still with her when she closed. The opening of the Metropolitan Opera was heralded annually at Justine's.

– Sue Putnam
The Memphis Flyer

Dining in Memphis calls for Justine's

This establishment, at 919 Coward Place, is conceivably the best restaurant in the South. The food is excellent.

In the main, the cooking of Justine's is French with Creole overtones. The New Orleans influence is altogether apparent with the likes of Shrimp Remoulade and Eggs Sardou, and to one mind, the restaurant has improved on the originals.

– Craig Claiborne
The New York Times
Wednesday, April 6, 1966

Hall arrangement on the commode

The South of moonlight and magnolias...

The history of Memphis is reckoned from the visit of DeSoto to the bluff villages of the Chickasaw Indians in 1541. This is one of several places where he is credited with discovering the Mississippi River. French explorers stopped by in 1673 and in 1682, LaSalle built a fort here. From the 1780's, the United States left the area in Indian hands until 1818, when Andrew Jackson, John Overton, and James Winchester laid out plans for a town which they named for Egypt's Memphis on the Nile, "place of good abode."

– Town and Country
June, 1961

Japanese lanterns hung throughout the garden.

How long since you tried something different for dinner at Justine's?

Prices at Justine's remain some of the lowest in town, incredibly so, for such quality.

– Edwin Howard
Memphis Business Journal
Feb 9, 1987

Secrets from the South's Best Restaurants

by Earlyne S. Levitas, published by Ballantine Books, New York.

Perhaps the single restaurant in the South that reigns as "the toast of its town" is Justine's. Nothing was spared by the owners in turning this graceful private home into a first-rate restaurant.

A gastronome rarely visits Memphis without trying one of the fabulous fish entrees – stuffed pompano, trout Marguery, or trout Amandine.

The Oriental Room – upstairs

Old oak floors, carefully brought back to life by refinishing, form the background for the Oriental Room.

– *Bruce Floor Wax Magazine*
Fall, 1959

JUSTINE'S

What can I say about Justine's that hasn't been said already? I can't even say anything that I haven't said already. Justine's has been in business for over thirty years, has won every national award worth sneezing over, and is a tangible credit to the community. It should be subsidized by the Arts Council as a civic treasure. The building and grounds are gorgeous and it would be a delightful place to eat and entertain if they served nothing better than airline food. The fact that the food is superb is almost too much to believe.

Any place that does that much business is bound to have good days and bad days, but I have never been there on a bad one. Private parties large and small, or just for dinner – it's always been great. The pavillions outside are the loveliest place in town for an outdoor party: not only are the gardens beautiful, but you also get to look all evening at the outside of the house, always a treat.

Everything on the menu, which is extensive, is good. You can't make a mistake. But if you don't know that Justine's gets better meat than anyone in town, please pay attention to me now. The filet, ordinarily about as interesting to me as iceberg lettuce, is something fine at Justine's. I don't know where or how long the beef is aged to aquire that taste, but you can't buy anything like it in the stores. With either fresh mushrooms or with Bernaise sauce, it is something special. I also love the sole Florentine – that is, if I can forego the crabmeat Justine. You get the point: it's all good.

Andy Hill/Memphis Magazine
(circa 1980's)

Silver punch bowl with our roses

Effulgent moon and scented night air give a story book mood...

Cotton Carnival began in 1872 when Memphis longed for an event that would bring cheer to the area weary of the Civil War, the recession, and a series of disastrous yellow-fever epidemics. For ten years, the city celebrated Mardi Gras with a New Orleans-type carnival. During the depression of 1931, Cotton had long ruled as king of the area. The couturier wardrobe began.

– *Town and Country*
June, 1961

There are 199 steps and 13 landings leading up to Justine's and Dayton's house in Vail, Colorado, and just as many steps leading to their phenomenal and abiding success in the restaurant business in Memphis. As the millennia are divided into B.C. and A.D., dining in Memphis is before and after Justine's.

[People had trouble finding us when they came to Memphis because we had no sign and not even a street number out front. We gave the cab drivers fits.]

But making Justine's hard to find probably helped make it become the "in" place, not only to dine, but to be seen.

The restored mansion was the final key.

The "Justinian Code" was complete.

And what is the "Justinian Code" for restaurant success: for today, she reduces its 199 steps to just three; elegant atmosphere, fine food, and meticulous service.

– Edwin Howard
"Life at the Top"
Memphis Business Journal
Dec. 12-16, 1983

Letters

I will always have wonderful memories of Justine's. It was such a welcome sight to see all the twinkling lights as the taxi entered Coward Place, knowing there would be an occasion for fine dining and friendship. Thank you and your staff for some of my happiest moments.

– J. Julian
Knoxville, Tennessee
January 22, 1996

I'll not only miss the wonderful food, but the good visits that I always enjoyed. My best regards to your staff, all of whom have been so helpful and thoughtful.

– Neil Block
Tunica, Mississippi
February 23, 1996

We had many, many years of "special occasions."

– John Nash
Memphis, Tennessee

Justines

Recipes

Escargots à la Bourguignonne

snails in garlic butter

- 12 T. garlic butter, see Sauces, Etc.
- 12 snails
- 12 shells or cups to cook in
- 12 circles puff pastry, if preferred; see Entrees

Originally, we used six snails to an order, baked in their shells and surrounded by rock salt. When the Health Department outlawed the shells, we used small porcelain cups topped with puff pastry.

Prepare the garlic butter.

Place snails in individual shells or cups and top each with a tablespoon of garlic butter. Bake in a preheated 450° oven for approximately 20 minutes. Serve hot and bubbling. Serves two .

If using porcelain cups, you may place a small circle of puff pastry on top of each butter-surrounded snail and bake until the crust is brown.

Serve with plenty of hot French bread and Beaujolais.

Oysters Rockefeller

baked oysters with spinach purée

- 2 nine-inch pie pans with rock salt
- 12 oysters on the half shell
- 12 T. Rockefeller sauce, see Sauces, Etc.
- 1/4 c. grated fresh Parmesan cheese

Fill pie pans with rock salt. Open oysters and cut away from the shell. Rinse under cold water. Place six shells, with their oysters, around each pan. Cover each oyster with Rockefeller sauce and sprinkle Parmesan over. Place in a preheated 450° oven and bake until cheese begins to brown, about 10 to 20 minutes. Serve very hot. Serves two.
Suggestion: This dish with a tossed salad and some hot French bread....

"Boogie" cracked the oysters every night. When he retired, Willie B. replaced him. Willie B. also took care of our gigantic vegetable and rose gardens and in the end, my father. On opening the oysters, Willie told me "I had to learn it the hard way." He advised me to wear a thick glove (he really told me "let a man do it") and to use an oyster knife. Run the knife between the oyster and the shell. Wedge it apart then run water over it to get the grit out. Set the shell, with an oyster in it, on a pan of rock salt. "Just like cleaning your nails. Just snap it open." Yeah, right. I am not going to try this at home. I miss Justine's.

Oysters Bienville

oysters in a crabmeat and mushroom crème sauce

Bienville is the only sauce that requires another sauce to make it.

Cover each raw oyster in its shell with Bienville sauce. Bake in a preheated 450° oven until sauce begins to brown, 10 to 20 minutes. Serve hot to two.

- 2 nine-inch pie pans with rock salt
- 12 oysters on the half shell
- 12 T. Bienville sauce, see Sauces, Etc.

> ——★
> *We always had pots of fresh sauces made up daily and on hand so, this dish is a little more work for you but well worth it.*

Oysters Casino

oysters in a pepper, celery and garlic sauce

Cover each raw oyster in its shell with Casino sauce. Bake in a preheated 450° oven until hot and bubbling, 10 to 20 minutes. Serve hot to two.

A Napa Valley Fume Blanc is heavenly with this.

- 2 nine-inch pie pans with rock salt
- 12 oysters on the half shell
- 12 T. Casino sauce, see Sauces, Etc.

Oysters Justine

oysters in a brown sauce
served in ramekin or fresh artichoke

Prepare Oysters Justine sauce. Mix flour, salt, black pepper and paprika together in a bowl. Lightly dredge oysters in flour. Place oil into frying pan and heat until hot. Add oysters and cook until crispy, about 5 minutes on each side. Drain on paper towels.

To serve: Divide oysters between two small ramekins and cover with Oysters Justine sauce. Put into hot oven and heat at 450° until bubbling. Sprinkle fresh parsley over the top. Serve piping hot. 2 servings.

or

Tuck 6 crisply sautéed oysters inside each artichoke and cover with Oysters Justine sauce. Sprinkle chopped parsley over top. Serve hot. Garnish plate with lemon and parsley.

- 1 3/4 c. Oysters Justine sauce, see Sauces, Etc.
- 1 c. flour
- 1 t. salt
- 1 t. black pepper
- 1 T. paprika
- 12 fresh oysters
- 1/2 c. vegetable oil or butter
- Either: 2 small ramekins or 1 larger one or 2 whole fresh artichokes, cooked with choke removed (place in hot water until ready for the oysters)
- 1 T. chopped fresh parsley

> ——★
> *Mom calls this one of her favorite dishes, and, with a sip of Clos Blanc de Vougeot; is one of the great taste combinations in dining.*

Oysters Two, Two and Two

Oysters Bienville, Casino and Rockefeller

Oysters on the Half Shell

We would put "snow" (crushed) ice on a pie pan, oysters in their shells around the edges with a dish of cocktail sauce (see Sauces, Etc.) in the center. Tabasco, horseradish and a wedge of lemon were served on the side.

We had the best and the fastest oyster openers there ever were. Outdoor cocktail parties would find customers standing around the open pit gobbling oysters down as fast as they appeared.

Shrimp Cocktail

fresh gulf shrimp on ice

Justine's served 6 to 8 large and very fresh shrimp in each cocktail order, placed over crushed ice, with cocktail sauce in the center, a sprig of fresh parsley and a lemon wedge on the side.

Crevettes Suprêmes

large shrimp cocktail

Some customers, who loved our shrimp so much, would order this as their entree. This pan held 16 to 18 fresh large shrimp.

Shrimp Rémoulade

cold shrimp with Rémoulade sauce

Prepared as shrimp cocktail substituting Rémoulade sauce or you may mix the sauce together with the shrimp and place on top of a bed of lettuce.

Crabmeat Cocktail

cold crabmeat cocktail

Arrange lettuce leaves on 2 plates. Rinse crabmeat under cold water and carefully remove any shell. Lightly pile the crabmeat "up" on top of the lettuce. Garnish plate with fresh parsley and lemon and serve with choice of sauce. Easy and delicious.

iceberg lettuce
8 oz. fresh jumbo lump crabmeat
Fresh parsley and lemon
Cocktail, Rémoulade or Hollandaise, see Sauces, Etc.

I didn't tell any of my friends I was going into business. I thought it was tacky. It was a year before most of my friends knew Justine of Justine's was me."
– Justine

Crabmeat Justine

hot crabmeat casserole

Put butter, sherry, Tabasco, Worcestershire and lemon juice together in a pan and simmer over low heat. When mixture is hot, add crabmeat and lightly fold over with a spoon, heat. Be careful not to burn or boil.

Put slices of toast in bottom of 3 small ramekins. Cover with drained crabmeat mixture (use a slotted spoon). Top with Hollandaise sauce and place in a preheated 450° oven. Bake until Hollandaise begins to brown. 8 to 10 minutes. Serve bubbling hot. 3 servings.

Words of Wisdom: Mom says, "Always fresh crabmeat, that's the secret." That's why she flew it in every day. "Season the butter, get it hot, heat crabmeat in butter before it goes into the oven so the Hollandaise won't burn before the crabmeat heats up. Drain the butter off with a slotted spoon."

6 T. butter
1/4 c. cooking sherry
dash of Tabasco
dash of lemon juice
dash of Worcestershire
1/2 lb. of the freshest possible lump crabmeat, rinse and remove pieces of shell
3 pieces French toast
3/4 c. Hollandaise, see Sauces, Etc.

No one ever got tired of Crabmeat Justine. I don't know why you couldn't substitute shrimp or lobster but everyone always wanted crabmeat.

Crabmeat Suzette

crabmeat in crêpe

- 10 oz. fresh lump crabmeat, rinse carefully removing any shell
- around 3/4 c. crème sauce, enough to "hold" the mixture together, see Sauces, Etc.
- dash of Worcestershire
- 1/2 t. lemon juice
- dash of Tabasco
- 1/2 c. grated yellow or mild cheddar cheese
- 4 crêpes, see below
- about 3/4 c. Hollandaise, enough to top each crêpe, see Sauces, Etc.
- lemon and parsley for garnishing

This dish accented the delicate flavor of the crabmeat.

In a sauce pan, add crabmeat, crème sauce, Tabasco, Worcestershire, lemon juice and grated cheese. Fold the mixture over low heat until it "holds" together. Remove from heat. Place crêpes on pan and divide the mixture between the 4. Roll up each crêpe.

When ready to serve, heat in a preheated 450° oven for 5 minutes. Remove and place on salad plate. Spoon Hollandaise sauce down the length of crêpe. Garnish plate with lemon and fresh parsley. Serves four. Serve with a Pouilly-Fuissé.

> *"Put your money into the food, not into the advertising."*
> *– Justine*

Crêpes

- 3 whole eggs
- 3/4 c. flour
- 1/2 c. milk
- 2 1/2 t. melted butter
- 1/2 t. salt
- 2 t. powdered sugar
- 1 c. half and half

In a bowl, with a wire whisk, beat eggs lightly. Gradually beat in flour, a little at a time. Add a little milk, if needed, to smooth sauce. Add melted butter, salt and powdered sugar. Beat in half and half with remaining milk. Beat for a few more seconds and set aside.

Heat non-stick sauté pan (the size you want your crêpes to be, thin crêpes!) and when hot, pour just enough mixture in to cover the bottom of the pan.

You are making these crêpes individually!

Turn pan slowly to cover completely and heat until crêpe is brown. Turn crêpe over and brown other side. Good luck.

I suggest you drink something stiff while trying this for the first time.

Artichaut de Vinaigrette, Hollandaise ou Beurre

fresh artichoke with Vinaigrette, Hollandaise or butter

With a sharp knife, slice inedible bottom of artichoke away. Carefully cut sharp points off leaves. Put water, rock salt and lemon in a pot with water level almost to the top of artichoke. Cover and simmer over medium heat for 30 minutes or until the leaves pull out easily. Remove from heat and cool. Drain upside down. Remove choke from center by opening up artichoke like a flower to reach the interior. Pull out tender center cone of leaves and scrape off exposed choke with a spoon. Good luck. Serve sauce on the side or inside of the artichoke.

2 fresh artichokes
1 1/2 qts. of water
1 T. rock salt or salt
half of a lemon
choice of dressing
Serve hot or cold with Vinaigrette, Hollandaise, or melted butter

Celeri de Roquefort

stuffed celery hearts

Mix eggs, bleu cheese, cayenne and mayonnaise together. Leave the leaves on the tender stalks and fill cavity with the mixture. Arrange celery on a tray, decorate with fresh parsley. Add carrot sticks and olives if preferred. Nice cold hors-d'oeuvre.

2 grated hard-boiled eggs
1 c. crumbled bleu cheese
dash cayenne
1/4 c. mayonnaise, see Sauces, Etc.
tender stalks of celery

Anchois

anchovies

Some customers liked them served on French toast. Phew!

"Guests dine in an aura of elegance and charm brought back from an era that helped foster the expression, "Southern Hospitality."
Bruce Floor Wax Magazine, Fall, 1959

8 oz. fresh chicken livers, rinsed
1 T. chopped yellow onion
1 t. salt
1/2 t. black pepper
3 c. water
1 T. butter
1 T. finely chopped green onion
2 t. finely chopped fresh parsley
1 t. dry mustard
dash of nutmeg
1/4 c. mayonnaise, see Sauces, Etc.
French toast
olives for garnish

Canapé Foie de Poulet

chicken liver pâté served with toast

Put livers, yellow onions, salt, black pepper and water together in a pot. Cover and simmer for about 10 to 15 minutes, until livers are tender. Remove from heat and strain. Purée mixture in a food processor.

Transfer to mixing bowl and add butter, stirring well. Add green onion, parsley, dry mustard, nutmeg and mayonnaise. Mix well.

Spread canapé on pieces of French toast. Garnish with sliced olives. May be put into a mold and refrigerated.

A delicious hors-d'ouevre.

Coeur d'Artichaut

artichoke hearts

Served with choice of salad dressing or Hollandaise.

Fruit Cup

assortment of fresh available fruit, cut into pieces

Frances Finnie, the salad chef who was there from the first day till the last, 48 years later, would arrange fresh orange sections, strawberries, etc...sprinkled with a touch of sugar.

Crevettes de Jonghe

shrimp sautéed in wine and vegetables

In food processor, put green onion, parsley and garlic together. Grind until fine. Add salt, cayenne, nutmeg and softened butter. Mix well.

Place shrimp in casserole and cover with ground mixture. Pour sherry over top. Mix bread crumbs and melted butter together and spread over top of shrimp. Place in a preheated 450° oven for 20 minutes or until bread crumbs brown and mixture is bubbly hot. Serves 2.

Delicious!

14 medium-sized fresh shrimp, cooked, peeled and deveined

3/4 c. green onion

1/2 c. parsley

1/2 clove garlic

1/4 t. salt

1/8 t. cayenne

10 T. soft butter

1/2 t. nutmeg

1 T. sherry

4 T. bread crumbs made from a loaf of French bread, toasted, crushed or rolled with a pin

2 T. melted butter

Wilma says, "A good cook is like a fine artist. They just use different tools."

Roties de Champignons Sous Cloche

broiled mushrooms with brandy sauce served on toast

Melt butter in pan over medium heat. When butter begins to brown, add mushroom caps, bottoms up. Spoon butter over mushrooms while cooking. When mushrooms begin to brown, turn over and cook 2 to 3 minutes longer. Pour in brandy. Carefully, and if at all possible, flame the dish for a few seconds.

Place toast on plate and put a mushroom on each piece. Pour sauce over the toast. Serves two or three.

6 T. butter

6 large mushroom caps, rinsed

1/4 c. brandy

6 pieces French toast

We would cover this dish to keep it hot all the way to the table.

Poulet Suzette

chicken in crêpe

(Or "Janet Suzette," because this is what I used to eat before I discovered the curry dishes. I really saved my parents money. I never ordered lobster or steak.)

12 ounces chicken, cooked and diced
1 c. crème sauce, see Sauces, Etc.
1 t. Worcestershire
dash of Tabasco
1/2 c. yellow or mild cheddar cheese, grated
4 crêpes, see Appetizers
3/4 c. Hollandaise, see Sauces, Etc.
parsley for garnish

Put chicken, crème sauce, Worcestershire, Tabasco and cheese together in a sauce pan. Stir over low heat until mixture holds together, remove from heat.

Place out 4 crêpes. Divide the mixture between each, spreading along the edge of the crêpe. Roll each crêpe over to form a roll. Place on a pan and heat in a preheated 450° oven for 5 minutes. Remove from oven and put on serving plate. Run Hollandaise sauce down the length of each crêpe and serve hot.

Mom always said she could burn water, but her taste buds were the best.

Soups and Salads

Potages

Soupe à l'Oignon Gratinée

French onion soup

- 8 T. butter
- 2 c. thinly sliced onions
- 1 T. flour
- 5 c. chicken broth, see Sauces, Etc.
- 2 t. Worcestershire sauce
- 1/2 t. Tabasco
- salt and pepper to taste
- 6 to 8 pieces of French toast
- 1/2 c. Parmesan cheese

In a pan, melt butter over medium heat. When butter begins to brown, add sliced onions. Sauté until onions begin to brown. Stir in 1 T. flour and add chicken broth, Worcestershire, Tabasco, salt and pepper.

Simmer over medium heat approximately 30 minutes. Do not boil. Float a piece of French toast over each serving. Sprinkle with freshly grated Parmesan. 6 to 8 servings.

Soupe de Cresson aux Pommes de Terre

watercress and potato soup

- 3 medium Idaho baking potatoes
- 1 medium onion, chopped
- 4 c. chicken broth, see Sauces, Etc.
- 1 T. chopped parsley
- 1/2 c. chopped watercress
- salt and pepper to taste
- 1 1/2 c. half and half

Peel and cut raw potatoes into small pieces. In a saucepan, put potatoes, onion, broth, parsley and watercress. Cook over medium heat until potatoes are tender, around 25 to 30 minutes.

Mash potatoes through a strainer or use a blender. Add salt and pepper to taste. Add half and half to your taste. Return to sauce pan and heat. Serve hot with chopped watercress sprinkled over the top. 6 servings.

Our watercress came from Huntsville, Alabama. Be sure it's fresh and dark green .

Vichyssoise

chilled potato and chive soup

3 medium Idaho baking potatoes
2 green onions, chopped
4 c. freshly-made chicken broth, see Sauces, Etc.
2 T. chopped fresh parsley
salt and white pepper to taste
1 1/4 c. heavy cream, chilled
A little fresh parsley to sprinkle over the top

Peel, dice and rinse raw potatoes. Place in a pot with onion, broth and parsley. Cook over medium heat until potatoes are tender. Put potato mixture in blender or mash through a strainer. Return to pot and add salt and pepper. Chill. To serve, stir in chilled cream and place in individual bowls topped with fresh parsley. Serves 6.

Senegalese

chilled curried Vichyssoise with chicken

3 medium Idaho potatoes
2 green onions, chopped
4 c. chicken broth, see Sauces, Etc.
2 T. chopped fresh parsley
salt and white pepper to taste
2 t. curry powder
1 1/4 c. heavy cream
2 T. cooked chicken, chopped
a little fresh parsley to sprinkle over the top

Peel and cut raw potatoes into small pieces. In a saucepan put potatoes, onion, broth and parsley. Cook over medium heat until potatoes are tender.

Mash potatoes through a stainer or use a blender. Return to pot and add salt, pepper and curry powder. Stir in cream until the soup thickens. Add chicken. Sprinkle fresh parsley over the top and serve cold. 6 servings.

Crème de Champignon

cream of mushroom soup

4 c. freshly-made chicken broth, see Sauces, Etc.
3/4 c. butter
3 c. chopped mushrooms
3 T. flour
salt and white pepper to taste
1 c. half and half

Simmer chicken broth over medium heat.

In sauté pan, melt 1/4 c. butter. Add mushrooms. Stir until darkened. Transfer to broth pot.

In sauté pan, melt 1/2 c. remaining butter. Stir in flour until smooth. Using a wire whisk, add flour mixture into the mushroom mixture. Add salt and pepper to your taste. Stir while cooking over low heat 15 minutes. Add cream to your desired thickness; you may not need the whole cup. Serve very hot. 6 servings.

SALADE

DAYTON SALAD

1 bunch of Bibb lettuce
2 T. hard-boiled egg, grated
2 T. crisp, freshly cooked, crumbled bacon
1/4 c. Roquefort dressing, see Sauces, Etc.

Cut off the bottom from the Bibb lettuce, rinse leaf by leaf under cold water. Drain in a colander, placing a towel over the top. Allow lettuce "to crisp" by placing ice on top of the towel and setting colander inside a bowl. Refrigerate until ready to use.

Place dry, crisp lettuce leaves in a bowl and toss gently with the Roquefort dressing. Place on salad plate, sprinkle with bacon and egg. Serves two.

My father's favorite that we named after him. An inside favorite of regular customers, not printed on the menu.

SALADE LAITUE AU ROQUEFORT

head of lettuce with Roquefort

2 lettuce squares, from a head of iceberg lettuce
1/4 c. Roquefort dressing, see Sauces, Etc.
2 squares of tomato

Remove tough outer leaves from lettuce and cut in half. Place on two plates and pour Roquefort dressing over the top. Garnish with tomato. Serves two.

SALADE AUX POINTES D'ASPERGES

asparagus tip salad

12 asparagus spears
iceberg lettuce leaves
1 quartered ripe tomato
2 T. salad dressing, Vinaigrette preferred

Rinse asparagus, snap off tough ends, and steam until tender, about 12 minutes. Chill. Divide a few lettuce leaves between two plates. Place asparagus on lettuce and cover with salad dressing. Garnish with tomato. Serves two.

Avocat Vinaigrette

avocado with Vinaigrette dressing

Peel avocado, cut in half, and remove the seed. Put lettuce on salad plate and place avocado on top. Put 1 T. of Vinaigrette in center of each half and garnish plate with parsley and tomato. Serves one or two.

- 1 ripe avocado
- 2 T. Vinaigrette dressing, see Sauces, Etc.
- fresh parsley and tomato for garnish
- 2 pieces of crisp lettuce leaves

Laitue à l'Huile et Vinaigre

Bibb lettuce with oil and vinegar

Rinse Bibb lettuce and chill. When ready to serve, toss lightly with oil and vinegar. Place iceberg lettuce leaves on plates and divide the Bibb salad between. Garnish with tomato wedge or a raw vegetable of your choice. Serves two.

- 1 bunch Bibb lettuce
- 1/2 c. oil and vinegar dressing, see Sauces, Etc.
- 4 crisp iceberg lettuce leaves
- 1 quartered tomato

Caesar Salad

prepared tableside

When a Caesar Salad was ordered, the waiter would go to the table with a tray holding wooden salt and pepper mills, chilled salad plates, cruets of oil and vinegar, small bowls of Parmesan cheese, croutons, egg and lemon. Garlic was "pressed" into a large wooden bowl and rubbed along the sides with a wooden salad spoon. An egg was cracked into the bowl then the Worcestershire was added. A lemon half was squirted in and stirred around. Oil and vinegar dressing was added, with the lettuce, and, from the looks of the wallpaper, flamboyantly tossed. Parmesan, anchovies and croutons topped the finished salad. Serves four.

We took it off the menu because it was ruining the wallpaper.

- 1/2 t. salt
- 1/4 t. ground black pepper
- 4 T. wine vinegar
- 6 T. olive oil
- 4 oz. Parmesan cheese, grated
- 1/4 c. croutons
- 1 raw egg
- half a lemon
- 1 clove garlic
- 4 anchovies
- 1 T. Worcestershire
- 4 c. Romaine lettuce, torn
- 2 c. iceberg lettuce

Croutons

We made our own croutons by cutting the French bread into little cubes and covering them with melted butter. The cubes were baked for 30 minutes in a 325° oven, turned once. When lightly brown and crunchy, they were ready to be cooled down and stored until used. Garlic or Parmesan could also be sprinkled over.

Epinard Salade

spinach salad

- 1 bunch fresh spinach
- oil and vinegar dressing, see Sauces, Etc.
- 2 T. crisp bacon, crumbled
- 2 hard-boiled eggs, chopped

Prepare spinach by removing most of the stem and rinsing thoroughly under cold water. Shake off the water and chill. When ready to serve, toss with dressing and serve topped with bacon and egg. Serves two.

Crabes Salade Laitue

crabmeat salad

- 8 leaves of Bibb lettuce
- 1/4 lb. fresh, jumbo lump crabmeat
- lemon and parsley to garnish
- choice of dressing: see Sauces Etc., our Thousand Island was preferred

Place lettuce leaves on plate. Rinse crabmeat under cold water, removing any shell. Stand it "up," in a peak, on top of the lettuce. Garnish plate with lemon wedge and fresh parsley. Serve with your choice of salad dressing. Serves two.

Cresson aux Oeufs et Lardons

watercress served with egg and bacon

- 1 1/2 bunches of fresh, dark green watercress
- 4 pieces of crisp iceberg lettuce leaves
- 1/4 c. oil and vinegar dressing, see Sauces, Etc.
- 2 T. crisp bacon, crumbled
- 2 T. grated hard-boiled egg

Remove long stems from watercress and rinse under cold water. Drain and chill.

Toss watercress with oil and vinegar. Place lettuce leaves on plate and "bunch" watercress up on top. Sprinkle bacon and egg over the top. Serves two.

My buddy, Edwin Howard's, favorite

Cresson à l'Huile et Vinaigre

watercress served with oil and vinegar

Cut off longer stems of watercress. Rinse under cold water. Place in strainer and cover with clean towel. Place ice on top and put in bowl. Chill to crisp the leaves.

When ready to serve, put leaves on salad plates. Put watercress in salad bowl and toss lightly with chilled oil and vinegar dressing. Stand the watercress up on the plates, over the lettuce. Serves two.

1 1/2 bunches of fresh watercress

1/4 c. oil and vinegar dressing, see Sauces, Etc.

4 pieces of crisp iceberg lettuce leaves

Salade Verte

tossed green salad

Prepare lettuce and spinach, then chill. Toss lightly with celery hearts and green onions. Arrange iceberg lettuce leaves on salad plates and divide the salad mixture. Stand the salad up attractively. Serve choice of salad dressing on the side or mixed in with the salad. Serves three.

4 c. of rinsed, cut or torn, fresh Romaine, iceberg and spinach leaves

1/4 c. chopped fresh celery hearts

1/4 c. chopped green onions

6 crisp iceberg lettuce leaves for plate

choice of salad dressing, see Sauces, Etc., French preferred

Mike Cannon, our pianist, remembered customers and their favorite songs.

Salade Mixte

combination salad

Rinse and chill the greens. Toss lightly with celery hearts, green onions, carrots and tomatoes. Cover plates with iceberg lettuce leaves and stand the salad up attractively. Add dressing or serve on the side. Serves two.

4 c. Romaine, iceberg, and spinach leaves (cut or torn)

1/4 c. chopped celery hearts

1/4 c. chopped green onions

1/4 c. grated carrots

1/4 c. diced tomatoes

6 crisp iceberg lettuce leaves for the plate

choice of dressing, see Sauces, Etc.

Salade de Tomates

sliced tomatoes on lettuce

- 1 nice ripe tomato
- 2 crisp iceberg lettuce, Romaine or escarole leaves
- fresh parsley for garnish
- choice of dressing, like Roquefort, oil and vinegar, our French, see Sauces, Etc.

Rinse and peel tomato, trimming away "anything that isn't red." Put lettuce on plate. Slice 4 tomato pieces and arrange on top of the leaves. Place a little parsley in the center and serve with your choice of freshly made dressing. Serves one.

Salade d'Anchois

combination salad with anchovies

- 4 c. rinsed, cut or torn, Romaine, iceberg and spinach leaves
- 1/4 c. chopped celery hearts
- 1/4 c. chopped green onions
- 1/4 c. grated carrots
- 1/4 c. diced tomatoes
- 1/2 c. oil and vinegar dressing, see Sauces, Etc.
- 9 strips of anchovies
- 6 crisp iceberg lettuce leaves

Prepare and chill lettuce leaves. When ready to serve, place leaves in bowl adding celery, green onion, carrots and tomatoes. Toss lightly. Pour the oil and vinegar over the salad mixture and toss. Divide salad among three plates and place anchovies on top. Serves three.

Dinner was served on burgundy and gold-rimmed dinner plates with the large iron gates engraved in the center.

Coeur d'Artichaut

hearts of artichokes with French dressing

- 4 crisp lettuce leaves
- 6 artichoke hearts
- 1/2 c. French dressing, see Sauces, Etc.
- tomato wedges
- parsley

Chill 2 salad plates. Arrange lettuce leaves on top of each, next artichoke hearts on lettuce and spoon dressing over. Garnish with tomato and parsley. Serves two.

Avocat à la Francaise

avocado with French dressing

Chill 2 salad plates. Arrange lettuce leaves on top of each, making a cup shape. Cut avocados in half and remove the seed. Peel if desired.

Put 2 halves of avocado on each plate. Pour French dressing into the center of each. Garnish plate with parsley and tomato. May make 4 smaller salads if preferred. Serves 2 to 4.

- 2 ripe avocados
- 1/4 c. French dressing, see Sauces, Etc.
- 4 crisp lettuce leaves
- parsley and tomato

Avocat Poulet

avocado stuffed with cubed breast of chicken

In a large bowl, mix chicken, diced egg, celery, sweet pickle, mayonnaise, pepper, salt and sugar. Arrange lettuce on salad plate. Put avocado half on top of each plate, filling with chicken mixture. Sprinkle paprika over the top. Garnish with hard-boiled eggs, tomato and parsley. Serves 2 and is fantastic!

- 2 c. cubed chicken, cooked
- 2 hard-boiled eggs, diced
- 1/2 c. chopped celery
- 1/2 c. diced sweet pickle
- 1/2 c. mayonnaise, see Sauces, Etc.
- 1/8 t. white pepper
- 1/4 t. salt
- 1 t. sugar
- 4 crisp leaves of iceberg lettuce
- 1 large ripe avocado, peeled, split in half, seed removed
- dash of paprika
- 1 hard-boiled egg, halved
- 2 tomato quarters
- parsley for garnish

My parents rarely dined out in Memphis, just danced.

Avocat Crabes

avocado stuffed with crabmeat

Arrange lettuce on 2 salad plates. Cut avocados in half, remove the seed and place on lettuce.

Lightly mix together the celery, onion, crabmeat and dressing. Divide between the plates, filling the center of each avocado. Allow some of the crabmeat to fall off to the side. Garnish with parsley, lemon wedge and tomato. To die for. Serves two.

- 4 iceberg lettuce leaves
- 2 ripe avocados, peeled
- 1/2 c. chopped celery hearts
- 1/4 c. chopped green onions
- 1 1/2 c. Thousand Island dressing, see Sauces, Etc.
- 2 1/2 c. fresh lump crabmeat
- parsley, lemon wedge, tomato wedge for garnish

Avocat Langouste

avocado stuffed with lobster

- 1 avocado
- 1 lemon
- 1/2 c. fresh lobster, cooked, chopped
- 2 t. lemon juice
- 1/4 c. chopped celery hearts
- 2 t. mayonnaise, see Sauces, Etc.
- 1 t. chopped fresh parsley
- crisp lettuce leaves
- lemon wedges and parsley for garnish

Cut avocado in half, peel and remove seed. Squeeze lemon juice on halves. In a bowl, toss, lightly together, the lobster, lemon juice, celery hearts, mayonnaise and parsley. Divide and stuff in the avocado halves placed on a bed of crisp lettuce. Garnish with lemon wedges and parsley sprigs. Serves one.

Not printed on the menu, but insiders knew

Salade des Crevettes

fresh shrimp salad

- 4 or 5 pieces of iceberg lettuce
- 1/2 c. chopped celery hearts
- 1/4 c. chopped green onions
- 1/2 c. diced, hard-boiled egg
- 1 1/2 c. Thousand Island dressing, see Sauces, Etc.
- 2 1/2 c. medium sized shrimp, cooked, split lengthwise
- a few dashes of paprika
- parsley, tomato quarters, hard-boiled egg and lemon wedges

Put lettuce on 2 plates. Mix celery hearts, green onions, diced egg dressing and shrimp together. Place on top of lettuce leaves. Sprinkle small amount of paprika on top of each. Garnish with parsley, tomato, egg and lemon wedge. Yummy! Serves two.

My Dad's interest in restorations led me to find the perfect subject matter from which to start a painting career.

Entrees

Entrees

Filet de Boeuf Béarnaise

filet with Béarnaise

4 8-oz. filets
1/2 c. Béarnaise sauce, see Sauces, Etc.
salt and pepper to taste

We grilled our filets on a charcoal broiler. If grill is not available, broil till done to preference. Season with salt and pepper. Top with Béarnaise. 4 servings.

Tournedos Béarnaise

filet with fresh artichoke bottoms and Béarnaise sauce

4 cooked fresh artichoke bottoms
4 8-oz. filets
1/2 c. Béarnaise sauce, see Sauces, Etc.
salt and pepper to taste

Heat artichoke bottoms in a small amount of water, butter and lemon juice. Cook filets, season with salt and pepper. Transfer to a warmed plate and place artichoke bottom alongside. Drizzle Béarnaise over each. Serves 4.

Filet de Boeuf aux Champignons Frais

filet with fresh mushrooms

4 T. of butter
2 c. rinsed and sliced fresh mushrooms
4 8-oz. filets
salt and pepper to taste

Heat butter in a sauce pan and allow to brown. Sauté mushrooms until darkened. Cook filets to individual preference. Top with hot mushroom mixture. Add salt and pepper. Serves 4.

We respectfully asked, on the menu, that the customers not smoke pipes or cigars in the restaurant, feeling it would interfere with the taste buds of the rest of the customers, but there were plenty of cigarettes smoked. Nino, who delivered the milk, said no one could ever quit smoking if they worked around Justine's.

Aloyau de Boeuf

sirloin strip steak

2 sirloin strips, cut about 1 1/2 inch thick

Salt and pepper to taste

Prepare charcoal grill. Cook strips 4 minutes on each side for rare, 6 minutes for medium. Season with salt and pepper. Always serve steaks on heated plates, which keeps the rarer steaks warm. May also cook steak in the oven, placing it on a pan and into a 475° preheated oven for preferred time.

Caneton Rovennais aux Cerises

roast duck with cherries

1 wild duck, split in half

2 pieces of celery

1/2 c. sliced carrots

1/2 c. sliced onions

1 t. salt

1/2 t. black pepper

3 qts. water

1 1/2 c. cooked wild rice, see Vegetables

12 dark bing cherries, with juice

1 1/2 c. brown gravy

parsley or watercress for garnish

<u>Fruit Glaze</u>

cherry juice

1/8 t. sugar

1 1/2 t. corn starch

water

<u>Gravy</u>

3 T. butter

1 T. flour

1 1/2 c. chicken broth

Place duck halves in a roasting pan. Add celery, carrots, onions, salt, pepper and water. Cover pan and bake in a preheated 375° oven for 2 1/2 to 3 hours, until duck is tender. Remove lid during the last half hour of baking. Prepare wild rice.

To serve: Place duck halves on 2 plates. Put large scoop of rice on each plate. Arrange cherries around the duck. Spoon glaze over the top of the fruit. Spoon brown gravy over duck and rice. Garnish with parsley or watercress. Serve hot. Serves 2.

To make fruit glaze: Mix the cherry juice with a touch of sugar. Bring to a boil. In a separate container, mix 1 1/2 t. corn starch with just enough water to dissolve. Stir into the fruit juice.

To make brown gravy: Melt 1/2 c. butter and allow to brown. Add 3 T. flour and stir until smooth. Add chicken broth to desired thickness.

This roast duck had to be ordered in advance. Guests would bring their own wild game by Justine's the day before their dinner.

Côtelette d'Agneau

lamb loin chops

4 center cut lamb chops
1/8 t. garlic
salt and pepper
3 T. mint jelly

Rub lamb chops with garlic, salt and pepper. Place on pan and cook in a preheated 450° oven about 12 minutes on each side for medium, less for rare. If using grill, grill for 10 to 12 minutes on each side, more or less, depending on thickness. We served 2 chops to each plate with a dish of mint jelly on the side.

Filet Mignon

char-broiled filet

2 filets
1/8 t. salt
1/8 t. black pepper
Béarnaise or mushroom sauce, see Sauces, Etc.

Prepare grill. Charcoal should be hot and smoking, not flaming. Put filets on grill and carefully cook each side 5 minutes for medium rare. Remove from heat, place on heated plate and serve. Serve plain or with Béarnaise or mushroom sauce on top or to the side. If grill not available, put steaks on a pan and broil in a preheated 450° oven for about 10 minutes. Season with salt and pepper after cooking to keep the juices in.

Coeur de Boeuf Bordelaise

hearts of beef tenderloin with Bordelaise sauce

4 small pieces of tenderloin, about 3 oz. each
3 T. Bordelaise sauce, see Sauces, Etc.

Place pieces of beef on skewers and grill for 5 minutes on each side. If grill not available, broil in a preheated 450° oven for 15 minutes or done to preference. Remove from heat and serve on a warmed plate. Top with Bordelaise sauce. Garnish plate with parsley. Serves 2.

With all the fresh sauces on hand, the cooks would sometimes top the Bordelaise with Béarnaise if the customers requested. Those were the days of butter, when no one knew the repercussions or cared.

Foie de Veau a Lardon

broiled calf liver with bacon

1/2 c. flour
1 t. salt
1/2 t. black pepper
8 slices of calf liver
1/4 lb. butter, melted
8 slices of bacon

Mix flour, salt and pepper together and put into a flat container. Cover each side of the liver with the seasoned flour. On a separate pan, place liver pieces flat and side by side. Spoon melted butter over the top. Bake in 400° preheated oven 10 to 12 minutes or less on each side, depending on thickness. May also sauté liver on top of the stove.

In a separate skillet, cook bacon. Serve liver hot with 2 slices of bacon on top of each. Serves 4.

3 T. butter
1 1/2 c. sliced onion
sprinkle of paprika

May also serve with smothered onions over the top. To prepare onions, melt 3 T. butter in sauté pan and add onions. Sprinkle with paprika and cook until transparent. Place onions on top of liver before adding the bacon. Delicious!

Beef Wellington

whole beef tenderloin baked in pâté and puff pastry

1 whole beef tenderloin, trimmed
1 recipe of puff pastry, recipe follows (may buy prepared pastry sheets)
1/2 c. liver pâté, see recipe
iced water
1 T. butter
2 whole eggs

Prepare grill. When coals are ready, put tenderloin on grill. Sear the tenderloin a few minutes on each side to burn any extra fat off. If grill not available, sear in a hot skillet with butter on top of the stove or in a hot oven. Remove from heat. Allow to cool.

Roll out the puff pastry dough. Make dough long and wide enough to cover the tenderloin. Spread liver pâté over the rolled out pastry. Put tenderloin on top of dough, fold dough over meat. Dip fingers into ice water and seal dough around meat. Tenderloin should be completely covered with smooth, well-sealed dough.

Grease a long pan, to hold the tenderloin, with 1 T. butter. Turn the sealed side of the tenderloin over and place on the pan.

Beat eggs with a wire whisk and rub over puff pastry. Put tenderloin in a preheated 400° oven and cook about 45 minutes. The pastry should cook to a golden brown. Remove from heat. Slice and serve. Red Bordeaux time!

Puff Pastry

2 c. sweet butter
4 c. unsifted flour
1 piece wax paper
1 t. salt
1 1/2 c. water
well-floured cloth
tin foil

Work the butter into a brick shape. Spoon 3 T. of flour onto a large piece of wax paper. Coat butter with flour. Wrap butter in wax paper and refrigerate.

In large mixing bowl, put remaining flour. Using your fingers, shape a well in the center of the flour. Add the salt. Gradually, add water to the flour, in a circular motion, until dough is firm and slightly sticky. Knead dough 20 minutes or more. Don't worry about overworking dough. Occasionally, pound dough on table while kneading. Dip fingers into water, dabbing a few drops onto the dough, to prevent from hardening or drying out. Dough is worked sufficiently when it becomes smooth and elastic.

Shape dough into ball and put aside for 15 minutes.

After 15 minutes, place dough on a well-floured cloth. Cut a cross shape in the center. Roll out 4 "ears" from the "cross," leaving center a thick cushion. Remove butter from refrigerator and place in center of the cushion. Fold dough over the butter. Seal edges together, covering the butter. Wrap in foil and chill for 20 minutes.

On floured cloth, gently roll dough, as evenly as possible, into a 1/3 inch thick rectangle. Do not roll over the end of the dough until the dough is 18 inches long. Bring ends of dough to center. Fold to make 4 layers of dough. Wrap in wax paper. Chill 1 1/2 hours. Repeat this routine four times. On fourth time, chill the dough for 3 hours. Dough is now ready to use. Gasp!

If you do not want to go to the trouble, you may buy prepared puff pastry. If you make the pastry, make the full amount. It may be used for other dishes and freezes well. You can just pull off what you need.

I wonder how the kitchen ever made all of these dishes; small kitchen, not many eyes on the stove, a work of genius, timing, every night, master chefs, excellent cooks, what a gift. How fortunate we were.

Casserole des Crevettes Crabes Champignons

shrimp, crabmeat and mushroom casserole

Prepare the crème sauce, sauté the mushrooms and boil the shrimp. Preheat oven to 400°.

In a large bowl, lightly fold together crabmeat, mushrooms, crème sauce, shrimp, Tabasco, and Worcestershire.

Portion servings into 4 one-serving size oven proof dishes or into one large. Bake 25 minutes or until the edges start to bubble. Remove and serve hot. May also include lobster in this dish. Serves 4.

1 1/4 c. crème sauce, see Sauces, Etc.

1/2 c. sautéed sliced fresh mushrooms, drained

16 boiled, fresh shrimp, cleaned and split lengthwise

2 c. fresh lump crabmeat, rinsed carefully and shelled

1/4 t. Tabasco

1 t. Worcestershire

Pompano Claudet

pompano with chives and garlic

Preheat oven to 450°.

Place pompano filets on a baking sheet and top with lemon juice, salt, cayenne, paprika and 1/2 c. melted butter. Bake in the oven for 20-25 minutes. (Cooking: It can be baked in the oven or placed under the broiler for a crisper version.) Be careful not to dry the fish out.

While the pompano is in the oven, melt 1/4 c. butter in a saucepan. Sauté green onion, garlic and parsley 5 to 7 minutes.

To serve: spoon garlic butter over the top. Garnish plate and serve hot. 4 servings.

4 filets of fresh pompano, rinsed and patted dry

1 T. lemon juice

1/4 t. salt

1/4 t. cayenne

1/4 t. paprika

1/2 c. melted butter

1/4 c. butter

1/2 c. chopped green onion

1 clove of garlic, chopped

1/4 c. chopped parsley

parsley and lemon garnish

Pompano Amandine

pompano with sautéed almonds

4 pieces of pompano, rinsed under cold water, patted dry
1/4 t. salt
1/4 t. cayenne
1 T. lemon juice
1/4 t. paprika
3/4 c. melted butter
1 stick of butter
1 c. sliced almonds
lemon and parsley for garnish

Place pompano filets on baking sheet. Sprinkle with salt, cayenne, lemon juice, and paprika. Spoon melted butter over the top. Bake in a preheated 400° oven for 20 to 25 minutes or until the edges start to brown.

While pompano is cooking, melt a stick of butter and allow to brown. Add the almonds, stir and allow to brown. Remove from heat.

Serve pompano, spooning almond mixture over the top. Garnish plates with lemon and parsley. Serves 4.

Pompano Grillé

pompano broiled in butter

4 fresh pieces of pompano
1/4 t. salt
1/4 t. cayenne
paprika to sprinkle over the top of fish
1/2 c. melted butter
3/4 stick of butter
1 T. lemon juice
lemon quarter and fresh parsley for garnish

Rinse pompano under cold water, drain and place on a baking sheet. Sprinkle with salt, cayenne and paprika. Spoon 1/2 c. melted butter over the top and bake in a preheated 400° oven for 20 to 25 minutes, until the edges begin to brown.

Prepare lemon butter. Melt 3/4 stick of butter in a sauté pan and allow to brown. Add 1 T. lemon juice.

Remove fish from oven, put on your serving platter or individual plates and spoon hot bubbling lemon butter over the top. Sprinkle with paprika. Serve hot. Garnish plate with lemon and parsley. Serves 4.

All of our parsley and lemons on the sides of the plates were just the perfect garnish for a rich meal.

Pompano en Papillote

pompano in a bag with oysters

- 4 pieces of pompano
- 1/2 t. salt
- 1/2 t. cayenne
- 2 t. lemon juice
- 1/2 c. melted butter or margarine
- 1 1/2 c. of claw crabmeat
- 3/4 c. chopped, boiled, and peeled shrimp
- 1/2 c. cooked, drained fresh mushrooms, sliced
- 3/4 c. Marguery sauce, see Sauces, Etc.
- 1/4 c. bread crumbs, see Sauces, Etc.
- 4 pieces of parchment paper
- 8 fresh oysters
- 2 t. butter

Rinse pompano under cold water, dry and place on baking sheet. Sprinkle with salt, cayenne and lemon juice. Spoon 1/2 c. melted butter over fish and place in a preheated 400° oven to partially bake for 15 minutes. Remove from oven and set aside.

In a bowl, put crabmeat, shrimp and mushrooms. Stir in the Marguery sauce. Carefully, add a few bread crumbs, a little at a time, being careful to not let the seafood become too dry.

Place the parchment paper out on the counter. Divide the seafood mixture, placing it in the center of each piece of paper. On top of each mound, place 2 oysters. Top with 1 piece of fish, placing a 1/2 t. butter on top of each.

Pull the paper together and fold until you "get down" to the fish. Fold ends over and turn folded side down on baking sheet. Rub leftover melted butter over the bags. Cover with another pan and bake in a 400° preheated oven for about 25 minutes or until the bags slightly brown.

Remove from the oven and put on serving plates. Cut the bag down the center and a little across sideways. Eat while very hot, from the bag. Serves 4.

——★ *A favorite*

"No effort or expense was spared by Justine and Dayton in recapturing the atmosphere of the traditional South of legend and fable, and incorporating it into their historic but highly modern restaurant. Wherever possible, the original doors, windows, shutters, cornices, balustrade, and woodwork of the antebellum structure have been restored to their original appearance."

– Bruce Floor Wax Magazine, Fall, 1959

Pompano Louisianne

pompano stuffed with crabmeat, shrimp and mushrooms

4 pompano filets
1 1/2 c. claw crabmeat
1 c. diced shrimp
3/4 c. sautéed, chopped mushrooms
3/4 c. Marguery sauce, see Sauces, Etc.
1/4 c. bread crumbs, see Sauces, Etc.
1/2 t. salt
1/4 t. cayenne
1 t. paprika
1 T. lemon juice
3/4 c. melted butter
1 1/2 c. Hollandaise, see Sauces, Etc.
1/8 t. Tabasco

Rinse the pompano filets under cold water. Slice a small piece of fish from the underside and save. Try to remove all of the bones.

In a bowl, mix the crabmeat, shrimp, mushrooms and Marguery sauce together. Add a few bread crumbs to hold the mixture together.

Fish will be cooked on a pan. Hold the fish in one hand. Use the other hand to place 1/4 of crab mixture on the fish. Place the small slice of fish, previously removed, on top of the stuffing. Turn fish over and place the stuffed side on the pan.

"Plump" the fish up. Sprinkle with salt, cayenne, paprika, lemon juice and add a touch of Tabasco. Spoon the melted butter over the top and bake in a preheated 450° oven for 25 or 30 minutes. Remove and place on individual plates. Cover with Hollandaise. Garnish plate with fresh parsley and lemons. Serves 4.

(May also use same recipe with trout or flounder.)

Flounder à la Meuniére

broiled flounder with a parsley-lemon butter

4 pieces of fresh flounder
1/2 t. salt
1/4 t. cayenne
1/4 t. paprika
1/2 c. melted butter
3/4 stick of butter
1 T. lemon juice
lemon quarters and fresh parsley for garnish
1 T. chopped fresh parsley

Rinse flounder under cold water, dry and place on a baking sheet. Sprinkle with salt, cayenne and paprika. Spoon melted butter over the top. Place in a preheated 450° oven. Bake for about 25 minutes, until the edges begin to brown.

In sauté pan, brown 3/4 stick of butter and add lemon juice. Place fish on plate and pour hot lemon butter over the top. Sprinkle parsley over and garnish the side of the plate. I ate this like crazy. Serves 4.

Or another version Wilma would make for me that's worth the whole cookbook just for me to have this recipe.... She would dust the sides of the fresh flounder with flour seasoned up with salt and pepper. On top of the stove, she would sauté the fish in oleo until the edges would begin to brown. It would be turned over once for thorough cooking then placed on a hot plate with just lemon squirted over the top. It was so good and I miss it so much. I miss the kitchen. I miss everyone I knew all of those years. It was all so much fun.

Truites Marguery

trout with shrimp crème sauce

4 pieces of fresh trout
1/2 t. salt
1/4 t. cayenne pepper
1 T. lemon juice
3/4 c. melted butter
1 3/4 c. Marguery sauce, see Sauces, Etc.
1 c. medium-sized, freshly cooked shrimp, peeled, rinsed and split
1/4 t. paprika

Rinse trout under cold water, dry and place on baking sheet. Sprinkle with salt, cayenne and lemon juice. Spoon the melted butter over the top and bake in a preheated 425° oven for approximately 25 minutes.

Put Marguery sauce and shrimp in a sauce pan and heat. Place fish on plate. Cover each piece with the sauce. Sprinkle with paprika. Serve hot. 4 servings.

Truites Florentine

trout and spinach casserole

4 trout filets
1/2 t. salt
1/4 t. cayenne
1 T. lemon juice
3/4 c. melted butter
3 c. crème spinach, see Vegetables
1/4 c. Parmesan cheese
2 c. Hollandaise, see Sauces, Etc.

Rinse trout under cold water. Place filets on sheet pan and sprinkle with salt, cayenne and lemon juice. Spoon melted butter over the top and bake in a preheated oven at 425° for approximately 25 minutes. Remove from the oven and set aside.

Heat the crème spinach. Place the fish in a separate oven-proof casserole. Cover with hot crème spinach and sprinkle with Parmesan cheese. Cover with Hollandaise and cook until sauce browns or until the edges are bubbling. Serve very hot. Serves 4.

The lemon butter brings out the flavor of the fish. Use the best butter, (we used Forest Hill) and the freshest trout. Ours was flown in fresh, daily from New Orleans.

Truites Grillées

broiled trout

4 pieces trout
1/2 t. of salt
1/4 t. cayenne
1/2 t. paprika
1 stick of margarine
1/4 c. butter
2 t. lemon juice
lemon and parsley for garnish

Rinse trout under cold water, dry and place on a baking sheet. Sprinkle with salt, cayenne and paprika.

In a saucepan, melt margarine. Spoon margarine over and under the trout to keep from sticking to the pan. Place in a 400° preheated oven and bake approximately 25 minutes. Remove from oven and set aside.

Melt 1/4 cup of butter in a saucepan and allow to start to brown. Add lemon juice and immediately pour over the trout and serve. Garnish plate with parsley and lemon.

Truites Amandines

broiled trout topped with toasted almonds

4 pieces of fresh trout
1 t. lemon juice
1/4 t. paprika
1/8 t. cayenne
1/2 t. salt
1 c. melted margarine
3 T. butter
3/4 c. almonds
parsley and lemon

Rinse trout under cold water, dry and place on baking sheet. Add lemon juice and sprinkle with paprika, cayenne and salt. Spoon melted margarine over fish. Bake in a preheated 400° oven for 20 to 25 minutes.

During cooking, melt 3 T. butter in a small skillet over medium heat. When butter begins to brown, add almonds. Sauté until almonds begin to brown. Remove from heat.

Place trout on serving plates (we always had them hot) and divide the almond butter over the top. Garnish plate with parsley and lemon. Serve hot. 4 Servings.

Omelette de Crabes

crabmeat omelette

3 whole eggs
3 T. half and half
1/4 t. white pepper
1/4 t. salt
1/2 c. fresh lump crabmeat, rinsed and shelled
1/4 c. crème sauce, see Sauces, Etc.
1/8 c. melted butter
parsley and lemon

Place eggs, half and half, white pepper and salt in a bowl and beat, about 4 minutes, with a wire whisk. Set aside.

Put crabmeat in saucepan. Add crème sauce. Carefully, stir while heating, being careful not to tear up the crabmeat pieces. Heat well.

In a separate non-stick skillet, heat melted butter without browning. Pour egg mixture into the skillet and lightly roll around. When omelette begins to set, put crabmeat on top and fold the omelette over. Slide onto a plate and garnish side of plate with fresh parsley and lemon. One entree-size serving.

Crabes Amandines

crabmeat with sautéed almonds

1/2 stick of butter
2 c. fresh rinsed lump crabmeat
1/8 t. Tabasco
1 t. lemon juice
1/2 t. Worcestershire
3/4 c. sliced almonds
lemon and parsley

Melt 1/4 stick of the butter in a small skillet. Add the crabmeat, Tabasco, lemon juice and Worcestershire. Carefully fold over, trying not to tear up the larger pieces of the crabmeat.

In another small skillet, melt remaining 1/4 stick butter and allow to begin to brown. Add the almonds and stir until they begin to brown. Remove from the heat and keep stirring or almonds will continue to cook.

Put hot crabmeat on warmed plates and sprinkle hot almonds over the top. Garnish side of plate with lemon wedge and fresh parsley. 4 servings.

Crabes Florentines

crabmeat and spinach casserole

Divide 1/2 T. butter between two casserole dishes. Layer with crabmeat, then crème spinach. Top with Parmesan cheese. Bake in a preheated 425° oven until bubbly hot. Remove from oven and cover with Hollandaise sauce. Return to the oven and bake until casserole browns or until very hot. Serves 2.

8 oz. lump crabmeat, rinsed and shelled
1 c. crème spinach, see Vegetables
1/2 c. Hollandaise, see Sauces, Etc.
1/4 c. Parmesan cheese
1 T. butter

Crabes Newburg

crabmeat in crème sauce

In sauce pan, mix crabmeat, crème sauce, lemon juice and Tabasco. Simmer over low heat until very hot. Place toast on individual plates or in a casserole. Cover with sauce. Sprinkle paprika over the top. Serves 4.

2 c. lump crabmeat, rinsed and shelled
3 c. crème sauce, see Sauces Etc.
1/2 t. Tabasco
1 t. lemon juice
1/2 t. Tabasco
8 pieces French toast
1/4 t. paprika

Crabes au Gratin

crabmeat casserole with cheese

In a small skillet, sauté 1 T. butter, mushrooms, salt and pepper. Set aside.

In a separate bowl, mix crabmeat, lemon juice, Tabasco and mushroom mixture. Gradually stir in the crème sauce. Place in casserole. Top with cheese and bits of remaining tablespoon of butter.

Bake in 375° preheated oven until cheese is melted and bubbling around the edge. Serve very hot. Serves 4.

2 T. butter
3/4 c. sliced mushrooms
salt and pepper to taste
2 c. of lump crabmeat
1 t. lemon juice
1/8 t. Tabasco
1 1/2 c. crème sauce, see Sauces, Etc.
3/4 c. shredded cheddar cheese

Crevettes et Crabes Gratinées

shrimp and crabmeat with cheese

16 medium-sized shrimp, cooked, peeled, cleaned and split lengthwise
12 oz. fresh lump crabmeat, rinsed and picked clean
1 t. Worcestershire
1 t. lemon juice
1/8 t. Tabasco
1 1/2 c. crème sauce, see Sauces, Etc.
1 c. shredded cheddar cheese
2 T. butter

In a bowl, mix together shrimp, crabmeat, Worcestershire, lemon juice and Tabasco. Gradually fold in the crème sauce.

Put in individual ramekins or one large. Sprinkle top with cheese and bits of butter. Bake in a preheated oven at 400° for about 20 minutes or until cheese is melted and bubbling around the edge. Serves 4.

I am having so much fun going over all of these recipes but, I am also homesick.

Artichaut de Crabe

fresh artichoke stuffed with crabmeat

1/4 c. diced celery
1/2 c. chopped green onions
1 1/2 c. of lump crab meat, rinsed and carefully cleaned for shells
1/4 t. cayenne
3/4 c. Thousand Island dressing, see Sauces, Etc.
2 freshly cooked artichokes, choke removed
1/8 t. paprika

Slice inedible bottom of artichoke away. Cut sharp points off leaves. Put water, salt and lemon in pot with water level almost to top of artichoke. Bring to a boil, cover, then simmer until leaves pull out easily, about 30 minutes. Remove from heat, drain upside down and allow to cool. Remove choke.

Put celery, green onion, crabmeat and cayenne in a bowl. Carefully fold in the dressing.

Stand the artichoke up in a serving bowl. Gently, spoon the crabmeat mixture inside, positioning it up, toward the top, to stand up high. Sprinkle paprika over the top and garnish plate with fresh parsley and a wedge of lemon. 2 servings.

Writing these recipes down brings up memories of all the years I watched these people cook. What an experience.

Champignons de Crabes

fresh mushrooms stuffed with crabmeat served over tomatoes

- 3/4 lb. lump crabmeat, rinsed and shelled
- 1/2 c. crème sauce, see Sauces, Etc.
- 6 large fresh mushroom caps
- 1 T. butter, in pieces
- dash of Worcestershire
- 1/4 t. Tabasco
- 2 t. lemon juice
- 1/4 c. bread crumbs
- 1/2 t. paprika
- 1/2 c. melted margarine
- 1/2 c. flour
- 1/2 t. salt
- 1/2 t. black pepper
- 1/8 t. cayenne
- 6 slices of tomato
- lemon and parsley for garnish

Put crabmeat in a bowl and add just enough crème sauce to hold the mixture together.

Put mushroom caps on a pan. Combine butter, Worcestershire, Tabasco and lemon juice and place in caps.

Form the crabmeat mixture into six balls and sprinkle each with bread crumbs. Place the balls on top of the mushrooms and sprinkle with paprika. Spoon half of the melted margarine over top of mushrooms. Bake in a preheated 400° oven for about 25 minutes.

While stuffed mushrooms are cooking, mix flour, salt, pepper and cayenne together. Flour both sides of the tomatoes and put on a small pan. Cover tomatoes with the remaining 1/4 c. margarine. Place pan in oven alongside mushrooms for about 10 minutes. Serve by placing 3 tomato slices per plate. Top each with a mushroom. Garnish plate with lemon and parsley. 2 servings.

Langouste Grillée avec Beurre Fondu

broiled lobster with drawn butter

- 2 live lobsters
- large pot for steaming lobsters
- around 10 c. of water
- 1 t. paprika
- 1 T. butter, melted
- 1/2 c. butter
- 2 T. lemon juice

Heat water, in covered pot, over high heat. When water begins to boil, add lobsters and cover pot, cooking for about 15 minutes or until lobsters are completely red. Remove from the pot.

Remove claws from lobster's body and turn the stomach side down. Cut the lobster in half, all the way down the backside. Place on a baking sheet with the inside body together and the lobster's legs on the outside. Sprinkle the meat in the tail with a little paprika and spoon a little butter over the lobster meat. Crack the claws slightly and place on the pan with the lobster. Place in the oven and broil for about 5 minutes. Remove from the broiler. Fold two cloth napkins and place on serving plates. Put 1 lobster on top of each napkin.

Melt the 1/2 cup of butter. Add lemon juice and heat until hot. Pour into 2 cocktail cups. Place cups on the plates. Garnish plates with lemon, parsley, lobster crackers, (the metal appliances, not saltines) and cocktail forks. Serves 2.

Langouste Thermidor

stuffed lobster

- 2 cooked lobsters
- 1 c. shredded cheddar cheese
- 1/4 c. chopped green onions
- 1 T. chopped fresh parsley
- 1/4 c. pimentos, chopped
- 1/2 c. sautéed, chopped mushrooms
- 1/4 t. Tabasco
- 1/2 t. Worcestershire
- 1 c. crème sauce, see Sauces
- 1/4 c. bread crumbs
- 1 T. butter
- 1/4 t. Paprika

Use a sharp knife and carefully split the lobster in half. Remove the meat, starting with the claws, and following with the tail section. Rinse the meat under cold water. Rinse out the shell. Place shell on a baking sheet, with the center together, and decorate "with a little lobster finger (leg) sticking out on each side."

Sprinkle small amount of cheese into the lobster shell. Dice the lobster meat into medium-sized pieces.

In a bowl, mix green onions, parsley, pimentos, mushrooms, Tabasco, Worcestershire and lobster. Stir in enough of the crème sauce to hold the mixture together. Add the bread crumbs. Divide the mixture, standing it up in the shell. Place a little meat in the top part of the lobster and in the tail section.

Cover top with remaining cheese and bits of butter. Sprinkle with paprika. Broil in a preheated 400° oven for 20 to 25 minutes. Bake until lobster begins to brown on and around the shell.

Put napkins on plates. Remove lobster from oven. Carefully place the lobster on top of napkin. Serve HOT. 2 servings.

I remember watching how just preparing the ingredients for each dish, cooked to order, was painstakingly done by the kitchen staff. I remember them stirring and stirring and stirring the sauces. Attention to detail in every way, put real "heart and soul" into each dish. Meticulous.

Oeufs Sardou

artichoke bottoms, spinach and poached egg casserole

Heat artichoke bottoms in a small amount of lemon juice and butter. In a separate pan, heat the crème spinach.

Poach 4 eggs: In a sauce pan, heat, to a boil, 3 cups of water containing a small amount of butter and salt. Carefully crack the eggs into the boiling water, turn the heat down and allow to simmer for 3 to 4 minutes. Remove with slotted spoon.

In each ramekin, layer spinach, 2 artichoke hearts with poached eggs on top of each. Spoon Hollandaise on top of each egg, and sprinkle paprika over the top. Serve very hot. 2 servings.

- **4 artichoke bottoms**
- **enough lemon butter in which to heat artichoke bottoms**
- **1 1/2 c. crème spinach, see Vegetables**
- **4 eggs**
- **1/2 c. Hollandaise, see Sauces, Etc.**
- **1/2 t. paprika**

Author Shelby Foote's "light" favorite.

Poulet Florentine

chicken and spinach casserole

Put chicken in a casserole or ramekin and cover with crème spinach. Put in a preheated 375° oven and bake until hot.

When dish is hot, remove from the oven and sprinkle Parmesan over the top. Cover with Hollandaise sauce and return to the oven. Bake until Hollandaise browns or dish bubbles around the edges. Serve very hot. Serves 2.

- **2 c. cooked chicken, sliced**
- **2 c. crème spinach, see Vegetables**
- **1/4 c. Parmesan cheese, grated**
- **3/4 c. Hollandaise, see Sauces, Etc.**

Everyone wore their couturier dresses to Justine's. It was a glamorous time in the city which sadly, few restaurants still maintain. Mom joked that we kept Lacledes and Frances Wright in business.

Foie de Poulet de Champignons

chicken livers with mushrooms and rice

- 12 fresh chicken livers
- 1/4 c. flour
- 1/2 t. black pepper
- 1/2 t. salt
- 1 t. paprika
- 1/2 c. margarine
- 3 T. butter
- 1 T. flour
- 1 c. freshly made chicken broth, see Sauces, Etc.
- 6 large mushroom caps
- 2 large scoops of cooked white rice
- 1 T. chopped fresh parsley

Rinse chicken livers and set aside.

In a bowl, mix 1/4 c. of the flour, black pepper, salt and paprika. Add livers and cover with the flour. In a skillet, melt 1/2 c. margarine over medium heat until hot. Add floured livers and partially cook.

In another skillet, melt 3 T. butter and cook until almost brown. Stir in 1 T. c. flour, stir until smooth. Add chicken broth and continue to stir. Add mushrooms and chicken livers. Put scoops of hot rice in center of warm plates. Spoon the livers and mushrooms around the rice. Sprinkle parsley over the top. Serves two.

Poulet aux Champignons Gratiné

chopped chicken and mushrooms with cheese

- 1 c. chopped mushrooms
- 1 T. butter
- 2 c. chicken, cooked and chopped into medium-sized pieces.
- 1/4 t. black pepper
- 1/4 t. Tabasco
- 1/4 t. Worcestershire
- 1 1/2 c. crème sauce, see Sauces, Etc.
- 1 1/4 c. shredded cheddar cheese
- 2 T. butter pieces

Sauté mushrooms in 1 T. butter until darkened. Drain. Mix chicken, mushrooms, black pepper, Tabasco, and Worcestershire together in a bowl. Add the crème sauce, mixing in a little at a time. "Fold" the mixture together until everything is mixed together well. Put into a casserole or ramekins. Sprinkle the cheese on top and dot with pieces of the remaining butter. Bake in a preheated 400° oven for approximately 30 minutes. Casserole is ready when its edges are bubbly hot and the cheese is completely melted. Serves 3.

"Sunday night around seven-ish, Penelope entered the musical atmosphere of Justine's."
– Commercial Appeal

Poulet Justine's

chicken served with mushrooms, artichoke bottoms and Béarnaise sauce

- 1 chicken half
- 1/2 t. salt
- 1/2 t. black pepper
- 1/2 c. celery
- 1/2 c. sliced onion
- 1/2 c. carrots, sliced
- 4 c. water
- 2 fresh artichoke bottoms
- 2 mushroom caps, sautéed
- 1/4 c. mushroom sauce, see Sauces, Etc.
- 1/4 c. Béarnaise, see Sauces, Etc.
- Fresh parsley garnish

In a roasting pan, put chicken, salt, black pepper, celery, onion, carrots and water. Place in a preheated 400° oven and bake approximately an hour and 20 minutes. Remove from the oven and transfer to a hot plate.

Place artichoke hearts on the side of plate. Top with Béarnaise. Place sautéed mushrooms on top of chicken. Top with mushroom sauce. Cover with Béarnaise. Garnish with parsley and serve immediately. Serves 1. Scrumptious.

Customers pre-ordered this dish since it takes more time to prepare.

Poulet Amandine

sliced chicken with almonds over rice

- 12 oz. of cooked chicken, cut slices 1/2″ thick
- 1/2 c. chicken broth, see Sauces
- 2 T. chopped green onion
- 1/2 t. chopped parsley
- 2 T. sautéed mushrooms
- 8 oz. of cooked white or wild rice
- 4 T. butter
- 4 T. almond pieces

Think of this dish as a mold. After preparation, it will be turned over and placed on a plate.

Place chicken slices on the bottom of a flat casserole. Top with broth, onion, parsley and sautéed mushrooms. Cover with the rice. Cover casserole with foil.

Bake in a preheated 450° oven until chicken is piping hot. Remove from oven and set aside.

In a sauté pan, melt butter over medium heat and allow it to begin browning. Add almond pieces and stir until they brown. Remove from heat.

Carefully, turn casserole over onto a serving dish. The chicken should be on the top. Cover with the toasted almonds and butter. Serve hot. Serves two.

Chicken Curry
Crabmeat Curry
Shrimp Curry

Your Choice of:

3 c. diced, cooked chicken

-or-

2 c. fresh lump crabmeat, rinsed and shelled

-or-

20 medium-sized freshly cooked shrimp, split and deveined

2 1/2 c. curry sauce, see Sauces, Etc.

2 scoops of cooked white or wild rice

Condiments:

1/2 c. raisins

1/2 c. blanched, sliced almonds

1 c. chutney

1/2 c. grated, hard-boiled egg

1/2 c. shredded coconut

1/2 t. paprika

Put curry sauce and choice of chicken, crabmeat or shrimp in a sauce pan. Heat over low heat.

In another pan, prepare rice.

Set up a condiment tray: Put raisins, almonds, chutney, egg and coconut in small separate containers.

Heat serving plates. Put scoop of hot rice in center of each. Spoon hot curry around the rice. Sprinkle top with a little paprika. Serve very hot. Pass the condiments. Serves 2.

Note: This was my favorite dish. I ate it at least 5 nights a week. I drove everyone in the kitchen crazy. I never could get enough. It was a great, stand-up, one-dish meal I would savor before the customers would arrive. My motto was "you could eat every night at Justine's and never gain an ounce" because everything was so fresh and skillfully prepared. My Aunt Georgia, who lives in Florida, would make the best fresh chutney for us from the mango trees in her back yard.

Aunt Georgia and Burt

Vegetables

Légumes

In the summer, Mom loved to serve the customers the beans, okra, squash, cucumbers and tomatoes that were picked from our garden just hours earlier. The fresh garden flavor was unmistakable.

Asperges Fraîches au Beurre

fresh asparagus served with butter

- 16 medium-size, fresh asparagus spears
- 4 T. butter
- 2 T. lemon juice

Rinse asparagus thoroughly, removing any sand from under the tips and snap off the tough ends. Steam until crisp and tender, about 12 minutes. In saucepan, melt butter and add lemon juice. Top asparagus with the lemon butter. Serve hot. Serves two.

Brocolis Hollandaise

broccoli served with Hollandaise

- one bunch of fresh broccoli
- 1/2 t. salt
- 2 c. water
- 1/2 c. Hollandaise, see Sauces, Etc.

Use a paring knife to prepare broccoli. Peel away tough outer skin and rinse well. Separate stalks and split pieces in half or leave whole.

In saucepan, put broccoli, salt and water. Cover and boil, 12 to 15 minutes, until tender. Be careful not to overcook. Drain and top with Hollandaise. Serve hot. Serves two.

Crème d'Epinards

crème spinach

- 2 c. chopped spinach, cooked fresh, drained
- 3/4 c. crème sauce, see Sauces, Etc.
- 1/2 t. salt
- 1/4 t. black pepper
- 1/8 t. Tabasco
- 1 t. Worcestershire
- 1 T. Parmesan
- 3/4 c. Hollandaise, see Sauces, Etc.
- extra Parmesan to sprinkle over the top

Put spinach, crème sauce, salt, black pepper, Tabasco, Worcestershire and Parmesan together in a saucepan. Heat, over low heat, until spinach is hot.

Place spinach in serving casserole. Sprinkle Parmesan over and then cover with Hollandaise. Place in a preheated 450° oven. Bake until Hollandaise browns and dish bubbles around the edges. Remove and serve hot. Serves four.

Pomme de Terre Suprême

potato stuffed with seafood

- 2 large baking potatoes
- salt and pepper to taste
- 2 T. butter
- 3 T. cream
- 1/4 c. crème sauce, see Sauces, Etc.
- 2 T. chopped and sautéed mushrooms
- 8 fresh shrimp, cooked, peeled and split lengthwise
- 1/4 c. fresh crabmeat
- dash of Worcestershire
- dash of Tabasco
- 3/4 c. grated yellow or mild cheddar cheese

Our kitchen scrubbed many a potato and wrapped them in foil. I used to love wrapping them, "to help out," when I was a kid. When hollowing out your baked potato, be careful not to break up the shell. Bake potatoes at 450° for an hour.

Place the scooped out potato in a pan and add salt, pepper, 1 T. butter and cream. Whip until potatoes are smooth.

In a separate bowl, mix the crème sauce, mushrooms, shrimp, crabmeat, Worcestershire and Tabasco together and put into the cavity of the potatoes.

Divide the creamed potatoes and cover the seafood with. Cover with grated cheese and remaining 1 T. of butter. Place in a preheated 400° oven for about 15 minutes. Remove and serve hot. Serves two.

Was not printed on the menu but an inside favorite. May be used for a main course or vegetable.

Pomme de Purée au Gratinée

mashed potato with cheese or the "stuffed potato"

- 2 large baked potatoes
- salt and pepper to taste
- 1 T. butter
- 3 T. cream
- 3/4 c. grated cheddar cheese

Bake potatoes at 450° for an hour. Slice the top off of the potato and carefully scoop the insides out. Mix potato pulp with salt, pepper, butter, and cream. Whip until smooth and return to shell. Sprinkle cheese over top. Place in the hot preheated 400° oven and bake about 15 minutes, until cheese melts and lightly browns. Serves two.

Tomate Bourre avec Crème d'Epinards

tomato stuffed with spinach

- 4 ripe tomatoes
- 1 c. crème spinach, see Vegetables
- 1 T. Parmesan cheese
- 2 T. Hollandaise

Rinse tomato and slice bottom away. Remove half of the core and fill with crème spinach. Place on pan and bake in 350° preheated oven for 15 minutes. Remove and sprinkle cheese over the top. Return to the oven for 5 minutes. Serve hot with a touch of Hollandaise over the top. Serves 4.

Pomme de Terre Gratinée

potatoes with cheese

4 c. boiling potatoes, cooked, peeled and diced
1/4 c. all-purpose flour
1/4 t. black pepper
dash of cayenne
1 t. salt
2 c. milk
1/4 c. chopped onion
1/4 c. chopped celery
7 T. butter
2 c. cheddar cheese, grated

In a saucepan, melt 4 T. butter. Sauté chopped onion and celery in butter for 3-5 minutes. Add flour, stirring until smooth with a wire wisk. Add 1/2 c. cheddar cheese. Stir over medium heat until creamy. Remove from heat and stir in potatoes.

Mix all remaining ingredients together thoroughly. Transfer to a casserole. Cover with remaining grated cheese and butter bits.

Bake in a preheated 400° oven for 20 minutes. Dish is ready when edges brown and dish is bubbly hot. Serves six.

Petit Pois Francaise

green peas

2 c. small green peas
1 c. water
1/2 t. salt
2 T. butter
pinch of sugar

Simmer peas, water and salt together. Cook over medium heat for 15 minutes. Remove from heat, drain, add butter and a pinch of sugar. Serve hot. Serves two.

Haricots Verts Lyonnaise

green beans with almonds

3 c. French-cut green beans
1 1/2 c. water
1/2 t. salt
3 T. butter
4 T. almond pieces

Simmer green beans, water and salt together for 15 minutes.

In a separate pan, sauté butter and almonds together until lightly browned. Remove from heat and continue stirring (the almonds will still be cooking). Drain beans and served topped with sautéed almonds. Serves four.

Celeris braises

braised celery

To prepare celery: cut the ends off the stalks and pull out 3 or 4 ribs. Slice ribs in half, then slice halves vertically. Rinse well. Set aside.

Melt butter over medium to high heat. Add celery and cook for a few minutes on each side. Remove from the heat and remove the celery.

Transfer celery to a pot and add the chicken broth and pepper. Cook over medium heat for 30 to 40 minutes or until celery is tender. Serves 1.

2 stalks of celery or 2 large celery hearts
1/2 stick of margarine or butter
4 c. of chicken broth, see Sauces, Etc.
1/4 t. black pepper
salt to taste

Mr. Abe Plough's (Schering-Plough Corp. founder) favorite.

Courgettes (en saison)

squash when in season

Slice squash and simmer in slightly salted water until partially cooked, not soft, 12 to 15 minutes.

Fill buttered casserole in layers: sliced squash, raw onion rings, salt, pepper, dot with butter, top with grated cheese. Repeat until casserole is filled, ending with cheese on top. Bake in 400° oven for 15 to 20 minutes until bubbly. Do not overcook. Serves 4.

1 1/2 lb. fresh yellow squash; Mom liked the small ones
1 large onion, sliced paper thin
1/4 t. salt
1/8 t. pepper
1/4 stick butter
1/2 lb. grated mild cheddar or Parmesan cheese

The phones were always ringing at our house. Mom would forget something was on the stove and the fire alarm would go off. Dad found burned pots hidden in the guest bath tub.

Le Riz Sauvage

wild rice

(love that French)

- 1/2 c. uncooked wild rice (yields 3 c. cooked)
- 2 quarts of water
- 6 T. butter
- 1/2 c. chopped green onions
- 1/2 c. fresh mushrooms, rinsed and chopped
- 1/4 t. black pepper
- 1/4 t. salt

Pour rice into boiling water. Simmer over low heat for an hour, adding more water as needed. When rice is tender, remove from the heat and drain. Steam in colander over hot water to keep warm.

When ready to serve: in butter, sauté the green onions and mushrooms. Cook until mushrooms darken. Add wild rice, salt and pepper. Serve hot. Serves 6.

New Year's Eve Black-Eyed Peas

- 1 lb. black-eyed peas, dried
- 1 t. salt
- 2 qts. water
- 1/4 lb. ham hock
- 1 yellow onion, peeled
- ground black pepper to taste, after cooking

The kitchen would pour the peas onto a table and remove rocks and dirt or "anything that shouldn't be there." Heat peas and water to boiling, cover and simmer for an hour. Add ham hock, onion and salt, simmering until tender, about 45 more minutes. Add additional water if needed. 6 to 8 servings.

We served 15 to 20 pounds of black-eyed peas on New Year's Eve "on the house" so everyone would have good luck in the following year. New Year's Eve dinners were always special at Justine's where everyone in formal attire would choose from a menu selected from the favorites. My father, dressed in tails, would serve the black-eyed peas, in a Louis XIV silver supper dish, to all the customers, giving his "personal," and genuine touch.

Rich Road Lunch

- 4 zucchini, sliced diagonally
- 2 ripe tomatoes in pieces
- 1 onion, yellow or white
- 1/4 c. Parmesan cheese
- 1/8 t. sweetener, Sweet & Low
- 4 T. butter

Melt butter and sauté zucchini and onions until tender, 12 to 15 minutes. Add tomatoes, sprinkle in the sweetener and Parmesan. simmer together for ten minutes. Serves 4 to 6.

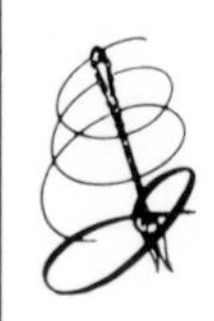

During the summers, Mom would sauté the fresh vegetables out of our garden for the neighborhood kids' lunch. She tried to teach us how to cook but we wanted to swim.

Desserts

Desserts

Garden mint or lotus ice cream homemade in hand-cranked freezers, was a memorable way to finish a meal unless one would rather have cheesecake topped with strawberries or rum cream pie.

If there was any room left, you could opt for cherries jubilee which were flamed at the table and served over vanilla ice cream, strawberries with kirsch, baked Alaska flamed with brandy served as a special occasion cake, unusual rum cream pie, our Dessert Justine, (blended vanilla ice cream and brandy), macaroon pie with vanilla or lotus ice cream, our own mint ice cream or ice cream with strawberries. The chocolate souffle was a favorite, the recipe had to be made for six people and ordered in advance, tying up an oven by itself during the night's business. And last, but not least, the chocolate mousse, a more or less cold version of the chocolate souffle. All delicious.

Cerises Jubilées

cherries flamed over vanilla ice cream

- 1 jigger of 100 proof brandy, see a liquor store
- 16 bing cherries, with some of their thick syrup
- 2 large scoops of vanilla ice cream

Heat a sauté pan and add the brandy. Light the brandy carefully (our waiters lit it flamboyantly) and add the cherries and syrup. Stir a bit, then spoon over ice cream. 2 servings.

Fraises au Kirsch

strawberries served with kirsch

- 14 "nice big red" strawberries
- 1/4 c. powdered sugar
- Jigger of kirsch, see liquor store

Remove tops of strawberries and rinse under cold water. Put 7 strawberries in 2 fancy glasses. Sprinkle with powdered sugar and pour kirsch over the top. Serves 2.

———★ *Simple, yet elegant.*

Gateau Alaska

baked Alaska flamed with brandy

20 slices of pound cake
1/2 gallon plain vanilla ice cream, in scoops
12 egg whites
1/2 t. cream of tartar
1 1/2 c. sugar
1 1/2 t. vanilla extract

On a large oven-proof tray, you will be layering cake and ice cream. Cake pieces will also cover the sides. A meringue will cover the cake and be browned in the oven.

To begin: Cover bottom of an oven proof tray with slices of pound cake. Cover slices with scoops of ice cream. Cover top and sides of ice cream with more slices of pound cake. Place in the freezer.

Put egg whites into mixing bowl. Add cream of tartar and beat until whites stand in peaks. Add sugar slowly. Add vanilla extract and fold the mixture over. Cover cake with egg white mixture smoothing it around. Use a spoon to make attractive swirls in the merinque. Bake in a preheated 450° oven for about 5 minutes, until egg whites lightly brown. Serve promptly. Serves about 10.

The waiters would flame this dessert, table-side, with a jigger of brandy. You do it carefully. Baked Alaskas were perfect for celebrating birthdays and anniversaries. Some customers called them Baked Alaskans.

Tarte a la Crème de Rhum

rum cream pie

For Crusts:
1/2 c. sugar
2 sticks of butter
2 c. graham crackers, crushed

For Filling:
7 egg yolks
1 c. sugar
1/2 c. water
2 envelopes of gelatin
1/2 c. rum, Bacardi dark
2 c. heavy cream
1 oz. piece of unsweetened chocolate

Prepare pie crusts: Mix sugar, butter and graham cracker crumbs together. Press mixture into two 9 inch pie pans. Bake the crusts in a 350° oven for a few minutes. Remove and cool. (You don't have to bake the crusts but some may prefer to.)

In a mixer, put egg yolks and sugar together. Beat until thick and lemon color. In a separate pan, mix water and gelatin together and stir over heat until dissolved. Stir rum and dissolved gelatin into egg mixture and set aside.

In a bowl, beat cream until stiff. Fold into egg mixture.

Divide mixture between 2 pie pans. Put into refrigerator until ready to serve. Grate chocolate over top of pies. 14 to 16 servings.

★ *Delicate and delicious!*

Dessert Justine

blended vanilla ice cream and brandy
the all-time favorite

Served, beginning in the early days, before "liquor by the drink." "Not that the customers ordered it for the liquor," Mom said, "because they usually hid that under the table in brown bags in those days."

- 6 scoops of plain vanilla ice cream
- 3/4 c. of brandy

Put ice cream and brandy in a blender. Blend until creamy or until the thickness of a milkshake. (Basically, it's 2 scoops to a jigger.) Put into freezer until ready to serve.

Makes 5 to 6 servings. I would flippantly and obnoxiously tell the customers, "It's an after-dinner drink, dessert and Alka-Seltzer all in one!"

This dessert would really finish off a full and rich meal, but watch out if you drink more than two. One was enough, two would get you tipsy and any more, you could not stand up. We served the "DJ" in a shallow champagne glass the customers would sip from.

Tarte de Macaron à la Mode

macaroon pie served with ice cream

- 2 c. sugar
- 1 c. chopped dates
- 2 t. baking powder
- 1 T. butter or margarine to grease pans
- 1 c. chopped pecans
- 1 c. cracker meal
- 7 large egg whites
- 1/8 t. of cream of tartar
- 1 t. almond extract

Mix sugar, dates, baking powder, pecans and cracker meal together in a bowl. Set aside.

In a mixer, begin to beat the egg whites and cream of tartar. Add the almond flavor. Beat until the whites stand up in peaks, then fold into the dry mixture.

Divide the mixture between two 9 inch buttered pie pans.

Bake pies in a preheated 350 degree oven approximately 30 to 40 minutes until golden brown. Remove from oven and cool. This recipe makes 14 to 16 servings.

We served the macaroon pie with a scoop of our home-made Lotus ice cream.

Glace de Menthe

mint ice cream

- 1 1/2 c. sugar
- 1 1/2 c. water
- 2 c. fresh mint leaves, rinsed
- 1/2 c. light corn syrup
- 2 c. pineapple, crushed
- 2 c. pineapple juice
- 2 c. milk
- 1/2 c. crème de menthe, visit a liquor store
- 1/4 c. lemon juice
- 2 c. heavy cream

In a saucepan, combine sugar and water. Let the mixture boil and cook to the "soft ball" stage (236°). Add the mint and cook ten minutes longer. Remove from heat and stir in corn syrup. Let cool.

Purée mixture in a blender, then strain into an ice cream container. Purée pineapple in a blender and transfer to the container. Add pineapple juice, milk, crème de menthe, heavy cream and lemon juice. Use a hand cranked or electric freezer and freeze until ready to serve. Approximately 10 to 12 servings.

My father brought the variety of mint we used down from Chicago when he first moved to Memphis. The variety was a hearty spearmint that my friends and I still keep growing in our yards, but it died in mine. I cannot lie.

Glace de Lotus

lotus ice cream

- 6 c. sugar
- 4 qt. half & half
- 2 c. lemon juice
- grated rind of 10 lemons
- 2 t. almond flavor
- 2 c. chopped blanched almonds

In container, put sugar, half & half, lemon juice, lemon rind and almond flavor. Mix together well.

Brown almonds, on a pan, in the oven. Cool and chop. Stir toasted pieces into the milk mixture and freeze. Keep ice cream in freezer until ready to serve. This makes about 18 scoops. Best served right after freezing. Light, lemony and wonderful! Refreshing and unusual. All of the above. Kemmons Wilson, founder of Holiday Inns, always requested the "soft ice cream made that day."

Soufflé au Chocolat

chocolate souffle

- 3 T. flour
- 1 1/2 c. milk
- 3 oz. unsweetened Baker's chocolate
- 1 t. vanilla extract
- 3 T. unsalted butter
- 6 egg yolks
- 1 T. butter for greasing bowl
- 1 T. sugar to sprinkle in bowl
- 6 egg whites at room temperature
- 3/4 c. sugar
- chocolate sauce, see below
- whipped cream
- 3 T. confectioner's sugar

In a bowl, stir the flour and milk together until smooth. Add the chocolate, vanilla extract and 3 T. butter. Place pan over boiling water and stir until chocolate melts.

In another bowl, beat egg yolks until thick and lemon colored. Stir into chocolate mixture. Stir over boiling water for 2 minutes. Set aside.

Thoroughly clean and dry a 3-quart souffle dish. Grease well with 1 T. of butter. Sprinkle with 1 T. sugar and set aside.

Pour the egg whites into a mixer and beat until almost stiff. Gradually add the 3/4 c. sugar and beat until stiff. Slowly fold the egg whites into the chocolate mixture, until no whites can be seen.

Put mixture into souffle dish and place in a pan of room temperature water. Bake in a preheated oven at 450° for about 5 minutes. Turn temperature down to 350° for an hour to an hour and 15 minutes. *Carefully* check appearance. Serve hot out of the oven. Sprinkle confectioners sugar over the top. Serve with chocolate sauce and whipped cream on the side. Serves 6.

> *Mom says if yours comes out like ours always did, you're a genius.*

Sauce au Chocolat

chocolate sauce

- 2 oz. unsweetened Baker's chocolate
- 2 c. milk (or half & half)
- 1 t. vanilla
- 1 3/4 c. sugar
- 4 oz. butter, not melted

Put chocolate, milk or half & half, vanilla and sugar in sauce pan. Stir over medium heat until sauce comes to a low boil. Simmer sauce until thick. Remove from heat, then add butter and allow to cool.

> *Advice to people who want to open a restaurant: "Don't...you have to be energetic and a workaholic and its tough."*
>
> *– Justine*

Gâteau à la Fromage Blanc de Fraises

strawberry cheese cake

I will always remember licking the spoon, from the cheesecake batter bowl, before taking it to the dishwasher, that "Apple" (Ethel Anderson, 'Chef Extraordinaire') would save for me. I could never decide which was better, the finished cake or the batter.

Filling:
- 5 eight oz. packages cream cheese
- 1 3/4 c. sugar
- 7 eggs
- 3/4 c. flour
- grated rind of 2 oranges
- grated rind of 6 lemons
- 2 t. lemon juice
- 1 t. vanilla extract

Crust:
- 1 1/4 c. flour
- 1/4 c. sugar
- grated rind of 2 lemons
- grated rind of 1 orange
- 1/2 c. margarine
- 1 egg yolk
- 1 t. vanilla extract
- 2 to 3 T. iced water
- A little flour to roll crust on

Topping:
- 1 c. sugar
- 1 c. water
- 1/2 c. chopped strawberries
- dash of red food coloring
- 3 T. cornstarch
- 6 – 7 c. water
- 25 large, rinsed stemless strawberries

Prepare the Filling: In a mixer, put cream cheese, sugar and eggs. Beat slowly until well mixed. Slowly add flour. Add orange rind, lemon rind, lemon juice and vanilla extract. Beat until creamy and smooth. Set aside.

Prepare the Crust: In another bowl, put flour, sugar, lemon rind, orange rind and margarine together, mix until crumbly. In center of the mixture, place the egg yolk and vanilla extract.

Using a fork, mix a small amount of ice water into the flour, a little bit at a time, until the mixture holds together, forming a soft ball, not sticky. Sprinkle a little flour on a board and roll the dough out. Fit dough into a 3 1/2 to 4 quart round casserole.

Pour the cream cheese mixture into the crust. Place in a preheated (475°) oven and bake until the cake begins to brown. Turn the oven down to 350° and bake until the center does not shake and a straw comes out dry, about 45 minutes. Remove from oven and cool.

Prepare the Topping: In a saucepan, put together sugar, 1 c. water, red food coloring and strawberries. Boil for a few minutes, then cook over low heat 10 minutes.

Stir cornstarch and water together until smooth. Gradually mix into strawberry mixture. Stir until thick.

Remove from heat and cover top of the baked cheesecake with the glaze. Cover the glaze with the large strawberries, top side down. Makes 10-12 servings.

Cheesecake is now ready to be served and is as pretty as a picture. Too bad the customers only saw the cut pieces; the whole cheesecake was a work of art.

Mousse au Chocolat

chocolate mousse

2 c. milk
2 packages Knox unflavored gelatin
6 oz. unsweetened Baker's chocolate
5 egg yolks
1 1/4 c. sugar
5 egg whites at room temperature
2 c. heavy cream

In a saucepan, over boiling water, cook milk, gelatin and chocolate. Stir until chocolate is melted.

Put egg yolks into mixer. Beat with 1 c. of sugar until thick and lemon-colored. Mix with chocolate mixture. Cook in a double boiler, for about 3 minutes. Remove and set aside.

Beat egg whites until almost stiff. Gradually add remaining sugar and beat until stiff.

Beat heavy cream until it stands in peaks. Fold egg whites and whipped cream into chocolate until no white is showing. Pour mixture into container and refrigerate until it sets up. Serve topped with chocolate sauce, and whipped cream.

La Glace aux Fraises

ice cream with fresh strawberries

1/2 c. sliced strawberries
1 T. sugar
2 dips of vanilla ice cream
5 whole strawberries
2 sprigs of fresh mint leaves

Mash sliced strawberries with sugar. Spoon over ice cream and top with whole strawberries. Garnish with sprigs of mint, if available. Serves 1.

Petite Merinques

meringue cookies
1 or 2 served à la mode

1/2 c. sugar
1/2 c. water
3 egg whites at room temperature
1/2 t. cream of tartar
1/8 t. salt
3 T. sugar
1 t. almond flavor

Put 1/2 c. sugar and 1/2 c. water in a sauce pan and boil until syrup thickens and has a shiny consistency.

In another bowl, beat egg whites with cream of tartar and salt until stiff. Beat in 3 T. of sugar. Add almond flavor. Slowly, beat in small amounts of the syrup mixture until all has been used. Place, by the spoonful, on a lightly greased cookie sheet. Bake in a preheated oven, at 300°, until dry and crackly. Makes approximately 3 dozen.

Sauces, Etc.

SAUCES, ETC.

Be patient, the secret to good sauces is slow cooking.

BÉARNAISE SAUCE

1 c. Hollandaise sauce, see Sauces, Etc.
1 t. Worcestershire
2 t. dried tarragon
2 drops Tabasco

Add tarragon, Worcestershire and Tabasco to Hollandaise sauce. This is for filets and Poulet Justine's.

BEEF BROTH

Trimmings from the beef tenders and racks of sirloins were used to make the beef broth. Sirloin made the best stock. From cutting the sirloins, the bones were put in a pot with celery, onions and carrots. The stock would simmer until it became rich, and then it would be strained to use for broth. If you don't want to go to the trouble, use a beef base.

BIENVILLE SAUCE

1/2 c. fresh claw crab-meat, finely chopped
1/2 c. fresh shrimp, cooked, finely chopped
1/4 c. bread crumbs, see Sauces, Etc.
3/4 c. Marguery sauce, see Sauces, Etc.
dash of Tabasco sauce
1 t. Worcestershire

Mix all ingredients together in a bowl. Chill until ready to use. For Oysters Bienville. (Would also be great on other baked seafood.)

The fresh, daily preparation of the sauces was the most important ingredient for the recipes. The batches they made were time consuming and always judged "just enough" by the reservations, so nothing was left over or saved overnight. Every dish would be as fresh as possible, the next day.... six days a week, 48 years....

Bordelaise Sauce

In a sauté pan, melt 2 T. butter over medium heat. Add mushrooms, green onions and parsley and cook until mushrooms darken. Stir to prevent sticking.

In a separate pan, melt 6 T. butter over medium heat. Stir in flour until smooth and golden brown. Add chicken broth and stir, using a wire whisk, until smooth. Add the cooked mushrooms, Worcestershire, Tabasco and sherry. Simmer over low heat for 20 minutes. Serve hot with steaks. Add salt and pepper to taste.

> *"Melt the butter with the flour till it's a golden brown, that's the color and the flavor."*

- 2 T. butter
- 2 c. sliced mushrooms
- 1/4 c. chopped green onions
- 1 T. chopped parsley
- 6 T. butter
- 3 T. flour
- 3 c. chicken broth, see Sauces, Etc.
- 3 t. Worcestershire
- dash of Tabasco
- 2 t. cooking sherry
- salt and pepper to taste

Bread Crumbs

The kitchen made plenty of fresh bread crumbs daily by toasting the loaves of French bread, crushing and rolling them with a pin. We used the Reising bread from New Orleans for 46 years.

Casino Sauce

In a food mill or grinder, put bell pepper, celery hearts, pimento, onions, parsley and garlic together and grind finely. Transfer to a saucepan and add butter, salt, and black pepper. Cook over low heat, stirring constantly.

Allow to simmer for 1 1/2 hours. Remove from heat and stir in bread crumbs. Cool and refrigerate until ready to use. Makes 5 cups. May freeze leftover. (Would also be great over shrimp.)

> *Mom disliked garlic and rarely used it.*

- 1 c. bell pepper
- 1 c. celery hearts
- 1 c. pimento
- 3/4 c. yellow onions
- 3/4 c. fresh parsley
- 1 clove of garlic
- 1/2 c. of butter
- 1 t. salt
- 1/2 t. black pepper
- 1/2 c. bread crumbs, see above

Chicken Broth

For 2 cups:
4 to 5 lbs. chicken, whole or cut into pieces
1 quart water or enough to cover chicken
1 celery stalk
1 small onion
1 carrot
1 T. salt
1 T. pepper

Put celery, onions, carrots and chicken in a stockpot. Season well with salt and pepper. Cover with water. Bring to a boil and cover pot. Simmer until the chicken is cooked. Chill and skim off the fat. Strain the broth and use for soups and gravies. If you are not cooking hens or chicken for meat and just need some broth, you can buy a chicken base or crystals and make whatever amount you need.

Hens were constantly being cooked in the kitchen so we always had plenty of fresh broth. Mom liked Tyson hens.

Cocktail Sauce

4 c. of tomato ketchup
1/4 c. grated fresh horseradish
1 t. lemon juice
dash of Tabasco
1 t. Worcestershire

Mix all ingredients together and chill until ready to use. Makes 4 cups. Serve with shrimp, oysters and crabmeat.

Boiled Shrimp

1 lb. shrimp
2 1/2 qts. boiling water
1 small onion
1 lemon sliced with rind
2 stalks celery
1/4 t. cayenne

The kitchen would cook 20 lbs. of shrimp at a time. I remember huge mounds of crushed ice covering the shrimp.

Combine all ingredients and boil for 6 to 8 minutes, until shrimp turn pink. Drain, rinse and shell.

Crème Sauce

1/4 c. all-purpose flour
1/8 t. white pepper
1/8 t. cayenne
1/2 t. salt
1/4 c. cooking sherry
2 egg yolks
3 1/2 c. half & half
1/4 c. mild cheddar
4 T. butter

In a saucepan, using a wire whisk, mix together flour, white pepper, cayenne, salt and egg yolks. Gradually stir in a small amount of cream into a smooth consistency. Add remaining cream, cheese and butter. Stir over heat until thick and creamy. Remove from heat and stir in sherry. Served with just about everything.

Curry Sauce

In a skillet, sauté butter, celery, and green onion together for 5 minutes. Turn heat off and add curry powder, flour, salt, white pepper, and cayenne. Stir together.

In separate saucepan, heat cream. Add the onion and celery mixture. Stir over heat until thick and creamy. Remove from heat and allow to cool.

May be refrigerated and used as needed. This is for crabmeat, shrimp, and chicken curries.

- 5 T. butter
- 1/4 c. finely chopped celery
- 1/4 finely chopped green onion
- 3/4 oz. curry powder (Mom loved Cross & Blackwell)
- 2 T. all-purpose flour
- 1/4 t. salt
- 1/8 t. white pepper
- 1/4 t. cayenne
- 2 1/2 c. half & half

These wonderful recipes were written from years of experience. The kitchen did not use much in the way of measuring cups. Most of it was guesswork. So relax and enjoy.

Garlic Butter

Process, finely together, the green onion, parsley, and garlic. Transfer mixture to a bowl and add the nutmeg, salt, cayenne and butter. Chill until ready to use. For the Escargot and Shrimp de Jonghe. May use with other baked seafood.

- 3/4 c. chopped green onions
- 1/2 c. fresh parsley
- 3 cloves fresh garlic
- 1/2 t. nutmeg
- 1/4 t. salt
- pinch of cayenne
- 12 T. soft salted butter

Hollandaise Sauce

Melt the butter. In separate bowl, whisk together egg yolks, cayenne and vinegar until yolks are fluffy and light. Add small amount of melted butter at a time, while beating, until all is used. For added thickness, place pan over boiling water and beat until desired consistency. Served with just about everything.

Our kitchen prepared everything from scratch and by hand. For an easier version, you could use a mixer and beat egg yolks, vinegar and cayenne together until thick and fluffy. Reduce mixer speed and slowly add the melted butter. Good luck. Makes 2 cups.

- 1 lb. butter
- 4 egg yolks
- 1/4 t. cayenne
- 4 t. cider vinegar

"Special" Salad Dressing

(A special version we used for private parties)

- 8 oz. cream cheese, softened to room temperature
- 1/2 c. chopped green onion
- 1/2 c. mayonnaise, see Sauces, Etc.
- 1 t. white pepper
- 1/2 c. tomato ketchup
- juice of 1 lemon
- 1 t. cayenne
- dash of paprika
- dash of Tabasco

Mix all ingredients well.

→ *When the regulars wanted something different.*

Marguery Sauce

- 4 T. butter
- 2 T. chopped green onions
- 1/2 c. mushrooms, chopped
- 1/4 c. flour, sifted
- 2 egg yolks
- 1/8 t. white pepper
- 1/2 t. salt
- 1/8 t. cayenne
- 3 1/2 c. half & half
- 1/4 c. mild cheddar cheese
- 1/4 c. sherry

Melt butter in a saucepan. Add green onions and mushrooms. Sauté until onions are transparent and mushrooms darken.

In a separate saucepan, combine flour, yolks, white pepper, salt and cayenne. Whisk in small amount of cream and stir until smooth. Add remaining cream, cheddar and the onion/mushroom mixture. Stir over heat until sauce is thick and creamy. Remove from heat and stir in sherry. Serve only with seafood.

Mayonnaise

- 1 1/2 T. prepared dry mustard
- 1/2 t. salt
- 1/2 t. sugar
- dash of cayenne
- 2 T. lemon juice
- 1 T. vinegar
- 3 whole eggs
- 3 c. chilled salad oil

In a small bowl, mix mustard, salt, sugar, cayenne, lemon juice and vinegar together.

In a mixing bowl, beat 3 eggs. Gradually beat in the mustard mixture. Beat in 1 c. of chilled oil. Slowly add remaining oil, in a steady stream, while beating. Refrigerate until ready to use. Makes 3 cups.

Mushroom Sauce

Saute 1 T. butter, green onion and mushrooms a few minutes, until mushrooms darken. Set aside.

In a separate saucepan, melt 5 T. butter. Gradually stir in flour. Stir until smooth. Add chicken broth, stirring constantly. Lower the heat. Add mushroom mixture, parsley, Worcestershire, Tabasco and sherry. Cook, on low heat, for 10 minutes. Use over chicken or filets.

- 1 T. butter
- 2 T. green onion
- 8 oz. fresh mushrooms, sliced
- 5 T. butter
- 1/4 c. all-purpose flour
- 4 c. chicken broth, see Sauces, Etc.
- 1 T. chopped parsley
- dash of Worcestershire
- 2 drops Tabasco
- 1/4 c. sherry

Oysters Justine Sauce

artichoke brown sauce

Melt 6 T. of butter in saucepan. Stir in 3 T. flour until smooth. Gradually add beef and chicken broth, stirring constantly. Lower heat. Add salt, pepper, Tabasco, sherry and Worcestershire.

In separate sauce pan, melt 2 T. butter, add chopped green onion, mushrooms, parsley and artichoke bottoms. Cook over low heat for 10 minutes, stirring until mushrooms cook. Combine with the previous sauce and simmer together for 10 minutes.

- 6 T. butter
- 3 T. flour
- 3/4 c. beef broth
- 3/4 c. chicken broth, see Sauces, Etc.
- 1/2 t. salt
- 1 t. ground black pepper
- 1/8 t. Tabasco sauce
- 1 T. cooking sherry
- 3 t. Worcestershire
- 2 T. butter
- 1/2 c. finely chopped green onions
- 3/4 c. finely chopped mushrooms
- 1 T. finely chopped fresh parsley
- 3/4 c. finely chopped artichoke bottoms

"Back in the 60's, for dining and entertainment; The Bell Tavern in the Claridge Hotel, The Summit Club at the top of the Sterick Building, The Embers Restaurant, and Justine's, an incomparable restaurant of international fame."

– Town and Country, 1982

Oil and Vinegar Dressing

- 1/2 c. wine vinegar
- 1 c. salad oil
- 1/4 t. salt
- 1/2 t. Equal dry sweetener
- 1 t. lemon juice
- 1/4 t. black pepper

Combine all ingredients and mix together well. Put in container with lid and shake. Refrigerate until ready to use. Always use one part vinegar to two parts oil. Toss lightly with salads. Makes 2 cups.

Rémoulade Sauce

- 3/4 c. celery hearts
- 3/4 c. green onion
- 1/4 c. parsley
- 1 1/2 c. mayonnaise
- 1/4 t. cayenne
- 1 t. Worcestershire
- 1 dash of Tabasco
- 1/4 t. salt
- 1 t. dry sweetener, Sweet & Low
- 2 t. lemon juice

In a food processor or grinder, finely chop celery hearts, green onions and parsley. Add remaining ingredients, mixing well. Keep chilled to use with seafood.

Rockefeller Sauce

- 1 stalk of celery
- 2 medium yellow onions
- 1/2 bunch fresh parsley
- 1 1/2 bunches of fresh spinach
- 1 c. butter
- 1 t. salt
- 1/2 t. black pepper
- 1 T. thyme
- 1/4 t. Tabasco
- 1 t. Worcestershire
- 3/4 c. bread crumbs, rolled or crushed from a baked crispy loaf of French bread
- 1/4 c. plus 1 T. Pernod, visit a liquor store (if you see Herbsaint while you're there, grab it)

Rinse celery, onions, parsley and spinach. Chop finely together in food processor. Melt 1/2 c. butter in a saucepan over low heat. Add the minced vegetables, salt and pepper. Cover and simmer approximately 1 1/2 hours. Add thyme, Tabasco and Worcestershire, mixing well. Cook an extra five minutes. Stir in bread crumbs and remaining butter. Add the Pernod last. Makes 6 cups. May freeze leftover. Our 5 gallons a day were always used.

In the afternoons, to make the Rockefeller sauce, the kitchen always started with a fresh bushel of spinach and plenty of fresh parsley. Some customers, who couldn't get enough of the sauce, ordered it for their vegetable. Our Rockefeller was originally made with Herbsaint which was later hard to find. (I had always heard that Herbsaint was an opium derivative. That's probably why the first customers fell in love with this dish.)

French Dressing

(ours was more Thousand-Island-ish, creamy, better)

Mix ingredients together in a bowl and chill. Serve with cold artichoke hearts, avocados and salads. Makes 3 cups.

- 1 1/2 c. mayonnaise, see Sauces, Etc.
- 1 1/4 c. ketchup
- 1/8 t. cayenne
- 1/4 t. sugar or Equal

Roquefort Dressing

Combine all ingredients and mix well. Refrigerate until ready to use. Makes 3 cups.

- 1 1/2 c. of crumbled bleu or Roquefort cheese
- 3/4 c. mayonnaise, see Sauces, Etc.
- 1/2 c. salad oil
- 1 t. lemon juice
- 1/4 t. cayenne

Thousand Island Dressing

In a bowl, mix together grated egg and olives. Stir in mayonnaise, ketchup, cayenne and sweetener. Mix together well. Chill until ready to use.

- 4 grated, hard-boiled eggs
- 3/4 c. of grated, stuffed green olives
- 2 c. mayonnaise, see Sauces, Etc.
- 1 c. tomato ketchup
- 1/4 t. cayenne
- 1/4 t. Sweet & Low

Vinaigrette

In a food mill or grinder, process celery, green onions, parsley, and pimentos. Grind into a fine mixture. Add vinegar, oil, Worcestershire, sweetener, salt and Tabasco. Mix well and chill until ready to use. Served with avocados, salads, and artichokes.

- 1/2 c. celery hearts
- 1/2 c. green onions
- 1/4 c. parsley
- 1/2 c. pimentos
- 1/4 c. red wine vinegar
- 3/4 c. salad oil
- 1 t. Sweet & Low
- 2 t. Worcestershire
- 1/4 t. salt
- dash of Tabasco

Wine List

We offered the best wines.....

Champagne

Dom Perignon

Mumm's Extra Dry

Moet & Chandon White Seal

Taittinger Brut LaFrancaise

Domaine Chandon, Napa Valley Brut

Burgundy Estate Wines

Richebourg, Frederick Wildman,

Bouchard - Gran Cru - 1977

Bonnes Mares, Frederick Wildman

La Tache

(Domaine de la Romanee-Conti)

Romanee Conti

(Domaine de la Romanee-Conti)

Eschezeaux, Bouchard Pere & Fils

Burgundy Red

Chevrey Chambertin, Louis Jadot

Fleurie, Louis Jadot

Pommard, Louis Jadot

Beaujolais Villages, Louis Jadot

Burgundy White

Puligny Montrachet, Louis Latour

Meursault, Louis Latour

Clos Blanc de Vougeot L'Heritier, Guyot

Chablis, Albert Pic

Beaujolais Blanc, Louis Jadot

Pouilly Fuisse, Bouchard Piers

Macon Blanc Villages, Bouchard Piers & Fils

De Ladoucette, Pouilly Fume

Bordeaux Red

Ginestet, Saint Emilion

Ginestet, Saint Julien

Chateau Lafite Rothschild 1964

Chateau Margaux 1969

Chateau Haut Brion 1969

Chateau Latour 1967

Chateau Montrose

Bordeaux White

Pavilion Blanc du Chateau Margaux, Ginestet

Chateau D'Yquem

Graves Extra, Ginestet

Chateau Voigny

California Red

Gamay Beaujolais, Beaulieu Vineyard

Pinot Noir, Beaulieu Vineyard

Cabernet Sauvignon, Charles Krug or

Beaulieu Vineyard

Georges de Latour, Private Reserve

California White

Pinot Chardonnay, Beaulieu Vineyard

Pinot Chardonnay, Monterey Vineyard

Chenin Blanc, Charles Krug or Mirassou

Fumé Blanc, Robert Mondavi

Chablis, Beaulieu Vineyard

Recipe Index

Order Form

Justine's
Memories & Recipes

Janet Stuart Smith
P.O. Box 11114
Memphis, TN 38111
Fax: 901-458-2306

Please send ________ copies of cookbook @ $21.95 each $ ______________

Shipping and Handling @ $ 3.00 each $ ______________

Tennessee residents add sales tax @ $ 1.81 each $ ______________

TOTAL $ ______________

Name ____

Address ____

City ____________________ State ____________ Zip ____________

Make checks payable to: *Janet Stuart Smith*

A portion of the proceeds from the sale of this book will go to MIFA (Memphis Inter-Faith Association) to help feed the hungry in the Mid-South.

Justine's
Memories & Recipes

Janet Stuart Smith
P.O. Box 11114
Memphis, TN 38111
Fax: 901-458-2306

Please send ________ copies of cookbook @ $21.95 each $ ______________

Shipping and Handling @ $ 3.00 each $ ______________

Tennessee residents add sales tax @ $ 1.81 each $ ______________

TOTAL $ ______________

Name ____

Address ____

City ____________________ State ____________ Zip ____________

Make checks payable to: *Janet Stuart Smith*

A portion of the proceeds from the sale of this book will go to MIFA (Memphis Inter-Faith Association) to help feed the hungry in the Mid-South.